# Advances of Spectrometric Techniques in Food Analysis and Food Authentication Implemented with Chemometrics

# Advances of Spectrometric Techniques in Food Analysis and Food Authentication Implemented with Chemometrics

Editor

**Ioannis K. Karabagias**

MDPI • Basel • Beijing • Wuhan • Barcelona • Belgrade • Manchester • Tokyo • Cluj • Tianjin

*Editor*
Ioannis K. Karabagias
University of Ioannina
Greece

*Editorial Office*
MDPI
St. Alban-Anlage 66
4052 Basel, Switzerland

This is a reprint of articles from the Special Issue published online in the open access journal *Foods* (ISSN 2304-8158) (available at: https://www.mdpi.com/journal/foods/special_issues/Spectrometric_Techniques_Food_Analysis_Food_Authentication_Implemented_Chemometrics).

For citation purposes, cite each article independently as indicated on the article page online and as indicated below:

LastName, A.A.; LastName, B.B.; LastName, C.C. Article Title. *Journal Name* **Year**, *Article Number*, Page Range.

**ISBN 978-3-03943-709-2 (Hbk)**
**ISBN 978-3-03943-710-8 (PDF)**

© 2020 by the authors. Articles in this book are Open Access and distributed under the Creative Commons Attribution (CC BY) license, which allows users to download, copy and build upon published articles, as long as the author and publisher are properly credited, which ensures maximum dissemination and a wider impact of our publications.

The book as a whole is distributed by MDPI under the terms and conditions of the Creative Commons license CC BY-NC-ND.

# Contents

About the Editor ............................................................. vii

**Ioannis K. Karabagias**
Advances of Spectrometric Techniques in Food Analysis and Food Authentication Implemented with Chemometrics
Reprinted from: *Foods* **2020**, *9*, 1550, doi:10.3390/foods9111550 ............................. 1

**Georgios Bekiaris, Dimitra Tagkouli, Georgios Koutrotsios, Nick Kalogeropoulos and Georgios I. Zervakis**
*Pleurotus* Mushrooms Content in Glucans and Ergosterol Assessed by ATR-FTIR Spectroscopy and Multivariate Analysis
Reprinted from: *Foods* **2020**, *9*, 535, doi:10.3390/foods9040535 ............................. 5

**Jiyu Peng, Weiyue Xie, Jiandong Jiang, Zhangfeng Zhao, Fei Zhou and Fei Liu**
Fast Quantification of Honey Adulteration with Laser-Induced Breakdown Spectroscopy and Chemometric Methods
Reprinted from: *Foods* **2020**, *9*, 341, doi:10.3390/foods9030341 ............................. 21

**Ivone M. C. Almeida, M. Teresa Oliva-Teles, Rita C. Alves, Joana Santos, Roberta S. Pinho, Suzene I. Silva, Cristina Delerue-Matos and M. Beatriz P. P. Oliveira**
Oilseeds from a Brazilian Semi-Arid Region: Edible Potential Regarding the Mineral Composition
Reprinted from: *Foods* **2020**, *9*, 229, doi:10.3390/foods9020229 ............................. 31

**Hongzhe Jiang, Fengna Cheng and Minghong Shi**
Rapid Identification and Visualization of Jowl Meat Adulteration in Pork Using Hyperspectral Imaging
Reprinted from: *Foods* **2020**, *9*, 154, doi:10.3390/foods9020154 ............................. 41

**Alex O. Okaru, Andreas Scharinger, Tabata Rajcic de Rezende, Jan Teipel, Thomas Kuballa, Stephan G. Walch and Dirk W. Lachenmeier**
Validation of a Quantitative Proton Nuclear Magnetic Resonance Spectroscopic Screening Method for Coffee Quality and Authenticity (NMR Coffee Screener)
Reprinted from: *Foods* **2020**, *9*, 47, doi:10.3390/foods9010047 ............................. 57

**Ewa Sikorska, Krzysztof Wójcicki, Wojciech Kozak, Anna Gliszczyńska-Świgło, Igor Khmelinskii, Tomasz Górecki, Francesco Caponio, Vito M. Paradiso, Carmine Summo and Antonella Pasqualone**
Front-Face Fluorescence Spectroscopy and Chemometrics for Quality Control of Cold-Pressed Rapeseed Oil During Storage
Reprinted from: *Foods* **2019**, *8*, 665, doi:10.3390/foods8120665 ............................. 69

**Ioannis K. Karabagias, Vassilios K. Karabagias and Anastasia V. Badeka**
The Honey Volatile Code: A Collective Study and Extended Version
Reprinted from: *Foods* **2019**, *8*, 508, doi:10.3390/foods8100508 ............................. 85

**Mariateresa Maldini, Gilda D'Urso, Giordana Pagliuca, Giacomo Luigi Petretto, Marzia Foddai, Francesca Romana Gallo, Giuseppina Multari, Donatella Caruso, Paola Montoro and Giorgio Pintore**
HPTLC-PCA Complementary to HRMS-PCA in the Case Study of *Arbutus unedo* Antioxidant Phenolic Profiling
Reprinted from: *Foods* **2019**, *8*, , doi:10.3390/foods8080294 ............................. 109

**Artemis Panormitis Louppis, Ioannis Konstantinos Karabagias, Chara Papastephanou and Anastasia Badeka**
Two-Way Characterization of Beekeepers' Honey According to Botanical Origin on the Basis of Mineral Content Analysis Using ICP-OES Implemented with Multiple Chemometric Tools
Reprinted from: *Foods* **2019**, *8*, 210, doi:10.3390/foods8060210 . . . . . . . . . . . . . . . . . . . . **123**

# About the Editor

**Ioannis K. Karabagias** (Ptychion of Chemistry, MSc. in Food Science and Technology, Ph.D. in Food Chemistry) is a postdoctoral researcher at the Laboratory of Food Chemistry in the Department of Chemistry at the University of Ioannina. His research interests include food chemistry, food analysis, food authentication, non-thermal methods of food preservation, food microbiology, food technology, natural antioxidants of agricultural products, the nutritional aspects of foods, and fermented foods.

*Editorial*

# Advances of Spectrometric Techniques in Food Analysis and Food Authentication Implemented with Chemometrics

**Ioannis K. Karabagias**

Laboratory of Food Chemistry, Department of Chemistry, University of Ioannina, 45110 Ioannina, Greece; ikaraba@uoi.gr; Tel.: +30-697-828-6866

Received: 24 September 2020; Accepted: 16 October 2020; Published: 27 October 2020

**Abstract:** Given the continuous consumer demand for products of high quality and specific origin, there is a great tendency for the application of multiple instrumental techniques for the complete characterization of foodstuffs or related natural products. Spectrometric techniques usually offer a full and rapid screenshot of products' composition and properties by the determination of specific bio-molecules such as sugars, minerals, polyphenols, volatile compounds, amino acids, organic acids, etc. The present special issue aimed firstly to enhance the advances of the application of spectrometric techniques such as gas chromatography coupled to mass spectrometry (GC-MS), inductively coupled plasma optical emission spectrometry (ICP-OES), isotope ratio mass spectrometry (IRMS), nuclear magnetic resonance (NMR), Raman spectroscopy, or any other spectrometric technique, in the analysis of foodstuffs such as meat, milk, cheese, potatoes, vegetables, fruits/fruit juices, honey, olive oil, chocolate, and other natural products. An additional goal was to fill the gap between food composition/food properties/natural products properties and food/natural products authenticity, using supervised and non-supervised chemometrics. Of the 18 submitted articles, nine were eventually published, providing new information to the field.

**Keywords:** *Arbutus unedo*; coffee; honey; meat; *Pleurotus* mushrooms; oilseeds; rapeseed oil; chemometrics

---

## General Remarks

Monitoring food and natural products quality, on the basis of highlighting specific properties of foodstuffs and natural based products, and at the same time detecting and preventing adulteration and fraud, is an important topic today at international level. Therefore, there is a challenging need for the development of rapid and validated analytical techniques such as gas chromatography coupled to mass spectrometry (GC-MS), inductively coupled plasma optical emission spectrometry (ICP-OES), isotope ratio mass spectrometry (IRMS), nuclear magnetic resonance (NMR), Raman spectroscopy, or any other spectrometric technique, for screening the authenticity and fraud of foods and natural products in combination with chemometrics, which offer the analyst the ability for the deep characterization of such matrices. The present special issue aimed to cover such demands. Let us go deeper through the published articles in this special issue.

The application of High-Performance Thin-Layer Chromatography (HPTLC) analysis and Liquid Chromatography High Resolution Mass Spectrometry (LC–HRMS), coupled with Principal Component Analysis (PCA) by Maldini et al. [1], was effective for the discrimination of *Arbutus unedo* plant material (leaves, yellow fruit, and red fruit) collected from La Maddalena and Sassari (Sardinia, Italy).The use of HPTLC and PCA comprised a simple and reliable untargeted approach to rapidly discriminate extracts based on tissues and/or geographical origins, while that of LC–HRMS in combination with PCA could be used as an effective approach for the identification of specific metabolites (quercetin, kaempferol, and myricetin derivatives, chlorogenic acid, arbutin, etc.) that werecapable to discriminate samples.

A validated $^1$H NMR spectroscopic method was implemented for the routine screening of coffee quality and authenticity by Okaru et al. [2]. A factorial experimental design was used to investigate the influence of the NMR device, extraction time, and origin of coffee on the content of specific metabolites such as caffeine, 16-O-methylcafestol (OMC), kahweol, furfuryl alcohol, and 5-hydroxymethylfurfural (HMF) determined in coffee. The aforementioned method was successfully validated for specificity, selectivity, sensitivity, and linearity of detector response. The proposed method produced a satisfactory precision for all analytes in roasted coffee, except for kahweol in canephora (robusta) coffee. The proposed validated method may be used for routine screening of roasted coffee quality and authenticity control (i.e., arabica/robusta discrimination), as its applicability was demonstrated during the recent OPSON VIII Europol-Interpol operation on coffee fraud control.

Sikorska et al. [3] managed to test the usability of fluorescence spectroscopy to evaluate the stability of cold-pressed rapeseed oil during storage. The quality deterioration of oils was evaluated on the basis of several chemical parameters (peroxide value, acid value, $K_{232}$ and $K_{270}$, polar compounds, tocopherols, carotenoids, pheophytins, oxygen concentration) and fluorescence, concerning the freshly-pressed rapeseed oil that was stored in colorless and green glass bottles, exposed to light and in darkness, for a period of 6 months. Parallel factor analysis (PARAFAC) of oil excitation-emission matrices revealed the presence of four fluorophores that showed different evolution throughout the storage period. The fluorescence study provided direct information about tocopherol and pheophytin degradation and revealed the formation of a new fluorescent product. Principal component analysis (PCA) applied on the analytical and fluorescence data showed that oxidation was more advanced in samples exposed to light due to the photo-induced processes; only a very minor effect of the bottle color was observed. Multiple linear regression (MLR) and partial least squares regression (PLSR) on the PARAFAC scores revealed a quantitative relationship between fluorescence and some of the chemical parameters (tocopherols and pheophytins).

Karabagias et al. [4] presented a research study that comprised the second part of a new theory related to honey authentication based on the implementation of the honey code and the use of chemometrics. Chestnut, citrus, clover, eucalyptus, fir, pine, and thyme honeys from Egypt, Greece, Morocco, Portugal, and Spain were subjected to gas chromatography coupled to mass spectrometry (GC-MS) analysis in combination with headspace solid-phase microextraction (HS-SPME). The application of classification and dimension reduction statistical techniques (multivariate analysis of variance, linear discriminant analysis, k-nearest neighbors, and factor analysis) were applied to the semi-quantitative data of volatile compounds. Results showed that honey samples could be distinguished effectively according to both botanical origin and the honey code ($p < 0.05$), with the use of hexanoic acid ethyl ester, heptanoic acid ethyl ester, octanoic acid ethyl ester, nonanoic acid ethyl ester, decanoic acid ethyl ester, dodecanoic acid ethyl ester, tetradecanoic acid ethyl ester, hexadecanoic acid ethyl ester, octanal, nonanal, decanal, lilac aldehyde C (isomer III), lilac aldehyde D (isomer IV), benzeneacetaldehyde, *alpha*-isophorone, 4-ketoisophorone, 2-hydroxyisophorone, geranyl acetone, 6-methyl-5-hepten-2-one, 1-(2-furanyl)-ethanone, octanol, decanol, nonanoic acid, pentanoic acid, 5-methyl-2-phenyl-hexenal, benzeneacetonitrile, nonane, and 5-methyl-4-nonene. New amendments in honey authentication and data handling procedures based on hierarchical classification strategies (HCSs) were exhaustively documented in the aforementioned study, supporting and flourishing the state of the art.

In the study of Louppis et al. [5], asfaka, fir, flower, forest flowers, and orange blossom honeys harvested in the wider area of Hellas by professional beekeepers were subjected to mineral content analysis using inductively coupled plasma optical emission spectrometry (ICP-OES). In total, 25 minerals were identified (Ag, Al, As, B, Ba, Be, Ca, Cd, Co, Cr, Cu, Fe, Hg, Mg, Mn, Mo, Ni, Pb, Sb, Se, Si, Ti, Tl, V, and Zn) and quantified. The mineral content varied significantly ($p < 0.05$) according to honey botanical origin, whereas lead, cadmium, and chromium contents ranged between 0.05–0.33 mg kg$^{-1}$, <0.05 mg kg$^{-1}$, and in the range of <0.12 to 0.39 mg kg$^{-1}$, respectively. Fir honeys from the Aitoloakarnania region showed the highest mineral content (182.13 ± 71.34 mg kg$^{-1}$), while flower honeys from Samos Island recorded the highest silicon content (16.08 ± 2.94 mg kg$^{-1}$). Implementation of multivariate analysis of variance, factor analysis, linear discriminant analysis, and stepwise discriminant analysis led to the perfect classification (100%) of these honeys according to botanical origin with the use of Al, As, Ca, Mg, Mn, Ni, Pb, Sb, Si, Zn and total mineral content. However, the higher lead content in the majority of samples than the recent regulated upper limit (0.10 mg kg$^{-1}$), sets the need for further improvements of the beekeepers' practices/strategies in honey production.

At the same time, honey adulteration comprises a major issue in food production, which may reduce the effective components in honey and, thus, has a detrimental effect on human health. The application of laser-induced breakdown spectroscopy (LIBS) combined with chemometrics was used by Peng et al. [6] for the rapid quantification of the adulterant content. Two common types of adulteration, including mixing acacia honey with high fructose corn syrup (HFCS) and rape honey, were quantified with univariate analysis and partial least squares regression (PLSR). In addition, the variable importance was tested with univariate analysis and feature selection methods (genetic algorithm (GA), variable importance in projection (VIP), selectivity ratio (SR)). The results indicated that emissions from Mg II 279.58, 280.30 nm, Mg I 285.25 nm, Ca II 393.37, 396.89 nm, Ca I 422.70 nm, Na I 589.03, 589.64 nm, and K I 766.57, 769.97 nm had compact relationship with the adulterant content. The most effective models for detecting the adulteration ratio of HFCS 55, HFCS 90, and rape honey were achieved by (SR-PLSR), (VIP-PLSR) and VIP-PLSR combined with root-mean-square error (RMSE) of 8.9%, 8.2%, and 4.8%, respectively. The study provided a fast and simple approach for detecting honey adulteration with cheap sweeteners.

In the study of Jiang et al. [7], a hyperspectral imaging (HSI) methodology was proposed to identify and visualize this kind of jowl meat adulteration in pork. Numerous hyperspectral images were acquired from adulterated meat samples in the range of 0–100% (w/w) at 10% increments using a visible and near-infrared (400–1000 nm) HSI system in reflectance mode. Mean spectra were extracted from the regions of interests (ROIs) and represented each sample accordingly. The performance comparison of established partial least square regression (PLSR) models showed that spectra pretreated by standard normal variate (SNV) performed best, with $R_p^2$ = 0.9549 and residual predictive deviation (RPD) = 4.54. Furthermore, functional wavelengths related to the identification of adulteration were individually selected using methods of principal component (PC) loadings, two-dimensional correlation spectroscopy (2D-COS), and regression coefficients (RC). The multispectral RC-PLSR model exhibited the most satisfactory results in prediction set that the $R_p^2$ was 0.9063, the RPD was 2.30, and the limit of detection (LOD) was 6.50%. Spatial distribution was visualized based on the preferred model, and adulteration levels were clearly discernible. The visualization was further verified that prediction results well matched the known distribution in samples. Overall, HSI was found to be a promising methodology for detecting and visualizing minced jowl meat in pork.

Furthermore, oilseeds from five native plant species with edible potential from the Brazilian Caatinga semi-arid region (*Diplopterys pubipetala*, *Barnebya harleyi*, *Croton adamantinus*, *Hippocratea volubilis*, and *Couroupita guianensis*) were subjected to mineral content analysis as reported in the study of M.C. Almeida et al. [8]. The minerals, Na, K, Ca, Mg, Fe, Cu, Cr, and Al, were analyzed by high-resolution continuum source atomic absorption spectrometry (HR–CS AAS) and P by the vanadomolybdophosphoric acid colorimetric method. The main elements found were K, Mg, and P (1.62–3.7 mg/g, 362–586 µg/g, and 224–499 µg/g on the basis of dry weight (dw), respectively). *B. Harleyi* seeds contained the highest amounts of K and P, while *C. guianensis* seeds were the richest in Mg. On the other hand, Fe was the most abundant oligoelement (2.3–25.6 µg/g dw). The Cr content was below the limit of quantification for all samples and Al content was low, ranging between 0.04–1.80 µg/g (dw). Linear discriminant analysis clearly differentiated *B. harleyi* and *C. guianensis* samples from the remaining ones. In sum, these oilseeds from the Brazilian Caatinga semi-arid region have the potential to be used as natural sources of minerals, mainly K.

Finally, attenuated total reflectance-Fourier transform infrared (ATR-FTIR) spectroscopy was used by Bekiaris et al. [9] to monitor the infrared absorption spectra of mushroom samples from *Pleurotus ostreatus*, *Pleurotus eryngii*, and *Pleurotus nebrodensis* strains, cultivated on wheat straw, grape marc and/or by-products of the olive industry. The spectroscopic analysis provided a chemical insight into the mushrooms examined, while qualitative and quantitative differences in regions related to proteins, phenolic compounds, and polysaccharides were monitored among the species and substrates studied. The use of advanced chemometrics, correlations of the recorded mushrooms' spectra versus their content in glucans and ergosterol, commonly determined through traditional analytical techniques, allowed the development of models predicting such contents with a good predictive power ($R^2$: 0.80–0.84) and accuracy (low root mean square error, low relative error, and representative to the predicted compounds spectral regions used for the calibrations). FTIR spectroscopy could then be exploited as a potential process analytical technology tool in the mushroom industry to characterize mushrooms and to assess their content in bioactive compounds.

**Funding:** This research received no external funding.

**Acknowledgments:** The author would like to thank the Foods MDPI Journal for selecting him as an academic Guest Editor. The contribution of articles by the Colleagues mentioned in the Editorial is greatly acknowledged.

**Conflicts of Interest:** The author declares no conflict of interest.

## References

1. Maldini, M.; D'Urso, G.; Pagliuca, G.; Petretto, G.L.; Foddai, M.; Gallo, F.R.; Multari, G.; Caruso, D.; Montoro, P.; Pintore, G. HPTLC-PCA Complementary to HRMS-PCA in the Case Study of *Arbutus unedo* Antioxidant Phenolic Profiling. *Foods* **2019**, *8*, 294. [CrossRef] [PubMed]
2. Okaru, A.O.; Scharinger, A.; Rajcic de Rezende, T.; Teipel, J.; Kuballa, T.; Walch, S.G.; Lachenmeier, D.W. Validation of a Quantitative Proton Nuclear Magnetic Resonance Spectroscopic Screening Method for Coffee Quality and Authenticity (NMR Coffee Screener). *Foods* **2020**, *9*, 47. [CrossRef] [PubMed]
3. Sikorska, E.; Wójcicki, K.; Kozak, W.; Gliszczyńska-Świgło, A.; Khmelinskii, I.; Górecki, T.; Caponio, F.; Paradiso, V.M.; Summo, C.; Pasqualone, A. Front-Face Fluorescence Spectroscopy and Chemometrics for Quality Control of Cold-Pressed Rapeseed Oil During Storage. *Foods* **2019**, *8*, 665.
4. Karabagias, I.K.; Karabagias, V.K.; Badeka, A.V. The Honey Volatile Code: A Collective Study and Extended Version. *Foods* **2019**, *8*, 508. [CrossRef] [PubMed]
5. Louppis, A.P.; Karabagias, I.K.; Papastephanou, C.; Badeka, A. Two-Way Characterization of Beekeepers' Honey According to Botanical Origin on the Basis of Mineral Content Analysis Using ICP-OES Implemented with Multiple Chemometric Tools. *Foods* **2019**, *8*, 210. [CrossRef] [PubMed]
6. Peng, J.; Xie, W.; Jiang, J.; Zhao, Z.; Zhou, F.; Liu, F. Fast Quantification of Honey Adulteration with Laser-Induced Breakdown Spectroscopy and Chemometric Methods. *Foods* **2020**, *9*, 341. [CrossRef] [PubMed]
7. Jiang, H.; Cheng, F.; Shi, M. Rapid Identification and Visualization of Jowl Meat Adulteration in Pork Using Hyperspectral Imaging. *Foods* **2020**, *9*, 154. [CrossRef] [PubMed]
8. Almeida, I.M.C.; Oliva-Teles, M.T.; Alves, R.C.; Santos, J.; Pinho, R.S.; Silva, S.I.; Delerue-Matos, C.; Oliveira, M.B.P.P. Oilseeds from a Brazilian Semi-Arid Region: Edible Potential Regarding the Mineral Composition. *Foods* **2020**, *9*, 229. [CrossRef] [PubMed]
9. Bekiaris, G.; Tagkouli, D.; Koutrotsios, G.; Kalogeropoulos, N.; Zervakis, G.I. *Pleurotus* Mushrooms Content in Glucans and Ergosterol Assessed by ATR-FTIR Spectroscopy and Multivariate Analysis. *Foods* **2020**, *9*, 535. [CrossRef] [PubMed]

**Publisher's Note:** MDPI stays neutral with regard to jurisdictional claims in published maps and institutional affiliations.

© 2020 by the author. Licensee MDPI, Basel, Switzerland. This article is an open access article distributed under the terms and conditions of the Creative Commons Attribution (CC BY) license (http://creativecommons.org/licenses/by/4.0/).

Article

# *Pleurotus* Mushrooms Content in Glucans and Ergosterol Assessed by ATR-FTIR Spectroscopy and Multivariate Analysis

Georgios Bekiaris [1], Dimitra Tagkouli [2], Georgios Koutrotsios [1], Nick Kalogeropoulos [2] and Georgios I. Zervakis [1,*]

[1] Laboratory of General and Agricultural Microbiology, Agricultural University of Athens, 11855 Athens, Greece; giorgosbekiaris@yahoo.gr (G.B.); georgioskoutrotsios@gmail.com (G.K.)
[2] Department of Nutrition and Dietetics, School of Health Science and Education, Harokopio University of Athens, 17676 Athens, Greece; d_tagkouli@yahoo.gr (D.T.); nickal@hua.gr (N.K.)
* Correspondence: zervakis@aua.gr; Tel.: +30-210-529-4341

Received: 25 March 2020; Accepted: 19 April 2020; Published: 24 April 2020

**Abstract:** Attenuated total reflectance-Fourier transform infrared (ATR-FTIR) spectroscopy was used to monitor the infrared absorption spectra of 79 mushroom samples from 29 *Pleurotus ostreatus*, *P. eryngii* and *P. nebrodensis* strains cultivated on wheat straw, grape marc and/or by-products of the olive industry. The spectroscopic analysis provided a chemical insight into the mushrooms examined, while qualitative and quantitative differences in regions related to proteins, phenolic compounds and polysaccharides were revealed among the species and substrates studied. Moreover, by using advanced chemometrics, correlations of the recorded mushrooms' spectra versus their content in glucans and ergosterol, commonly determined through traditional analytical techniques, allowed the development of models predicting such contents with a good predictive power ($R^2$: 0.80–0.84) and accuracy (low root mean square error, low relative error and representative to the predicted compounds spectral regions used for the calibrations). Findings indicate that FTIR spectroscopy could be exploited as a potential process analytical technology tool in the mushroom industry to characterize mushrooms and to assess their content in bioactive compounds.

**Keywords:** mushroom; *Pleurotus*; glucan; ergosterol; mid-infrared spectroscopy; FTIR; spectroscopy; chemometrics; prediction

---

## 1. Introduction

During the last two decades, there has been a 30-fold increase in the global supply of cultivated edible mushrooms, following their constantly increasing consumption [1]. *Pleurotus* mushrooms hold the second place in the total production, which is mostly due to their relative ease of cultivation in a wide range of lignocellulosic agro-residues combined with their rather limited infrastructure requirements [1–5]. *P. ostreatus* presents a cosmopolitan distribution and is the most widely cultivated *Pleurotus* species. However, during the last decade, *P. eryngii* ("king oyster") mushrooms demonstrated a steep increase in demand that is mainly attributed to their excellent organoleptic properties resulting in a 3–5 times higher selling prices in comparison with *P. ostreatus* [1]. Similarly, *P. nebrodensis* is also a choice edible mushroom species and the only fungus included in the Top 50 Mediterranean Island Plants [6]; therefore, its commercialization is of significant importance.

*Pleurotus* mushrooms are of significant nutritional value (i.e., relatively high content in proteins, vitamins and minerals, low amount of fats) which make them ideal for consumption by people suffering from hypertension, high blood low-density lipoprotein (LDL), cholesterol or triglycerides levels, obesity, metabolic diseases and diabetes [5]. Furthermore, they contain bioactive compounds associated with antitumor, antioxidant and immunomodulating activities [7]. Among them, β-glucans are high molecular weight constituents of fungal cell walls found also in *Pleurotus* species, which are linked with health-beneficial properties [8–10]. Ergosterol exists in fungal cell membranes and is a well-known precursor of vitamin $D_2$ [11]; *Pleurotus* mushrooms, in particular, demonstrate relatively higher concentrations and better conversion kinetics of ergosterol to vitamin $D_2$ in respect to other cultivated species [12]. Due to the significance of such bioactive compounds, recent studies focused on increasing their content, and therefore the nutraceutical properties of the mushrooms, by modifying/optimizing the production processes [13–15].

*Pleurotus* mushrooms are commonly cultivated on substrates composed of cereal straw supplemented with wheat-, rice- or soy-bran and/or flours from various leguminous seeds. In addition, many other locally abundant plant and food residues are also used for their large-scale production [5]. Among them, olive mill and winery by-products are two of the main agro-industrial wastes generated in the Mediterranean region, and their effective management and safe disposal are particular challenging due to their huge volume, seasonality of production and physicochemical characteristics (e.g., high phenol, lipid and organic acids content, acidic pH) [16,17]. However, both grape marc and olive mill residues contain organic compounds with bioactive properties [18,19]; therefore, their exploitation as substrates for the production of mushrooms with enhanced functionality is much sought-after, but also feasible, as it was recently demonstrated [2,10,13].

Traditional analytical techniques/assays used for measuring bioactive compounds in mushrooms can be laborious, time-consuming and expensive. Process analytical technology (PAT) provides valuable data about chemical processes to be used for monitoring and optimization purposes [20]. In PAT, the combination of appropriate measurement devices with multivariate statistical analysis (chemometrics) creates tools which can rapidly, accurately and usually non-destructively assess the quality, quantity and certain functional properties of various organic compounds [21]. Fourier transform infrared (FTIR) spectroscopy has been an ideal PAT tool for the food industry [22–25] since it can provide detailed information about the molecular structure of specific compounds of interest. Moreover, when combined with advanced chemometrics, it leads to the prediction of their content in the final product, thus allowing its implementation in the form of an on-line/at-line process analyzer [26]. In the mushroom industry, such a tool (i.e., predicting the content of mushrooms in selected constituents) could be of great interest both for the growers, as a way of promoting a product of high nutraceutical value, as well as for companies processing mushrooms to produce health promoting foods or drugs/cosmetics. To date, FTIR spectroscopy has been mainly applied to identify various filamentous fungi [27,28], to delimit taxa within the genera *Pleurotus*, *Ganoderma* and *Boletus* [29–31], to discriminate among mushroom samples of the same species on the basis of geographic origin [32] or to evaluate the post-harvest quality properties in *Agaricus bisporus* mushrooms [33]. To the best of the authors' knowledge, no PAT tool exists for the assessment of mushroom content in bioactive compounds.

In this study, attenuated total reflectance-Fourier transform infrared (ATR-FTIR) spectroscopy was applied to obtain a chemical insight into the *Pleurotus* mushrooms produced on various substrates, and to develop chemometric tools to accurately determine/predict their content in glucans and ergosterol.

## 2. Materials and Methods

*2.1. Biological Material*

Twenty-nine strains of *P. ostreatus* (#15), *P. eryngii* (#13) and *P. nebrodensis* (#1) were used in this study. All strains were routinely preserved on potato dextrose agar (PDA, Difco; Fischer Scientific, Hampton, NH, USA) and maintained in the Culture Collection of the Agricultural University of Athens, Laboratory of General and Agricultural Microbiology. Each strain was cultivated in up to three substrates, which resulted in a total of 79 samples used for the determination of glucans and ergosterol contents. Results on ergosterol content in the other 30 samples of *P. ostreatus*, *P. eryngii*, *P. nebrodensis* and *P. citrinopileatus*, obtained in previous experiments [13,14], were added to increase the model's variance during the calibration, achieving, in this way, a better prediction performance and accuracy.

### 2.2. Cultivation of Pleurotus Species

Three substrates, i.e., wheat straw (WS; control), grape marc plus wheat straw (GM; ratio 1:1 *w/w*) and two-phase olive mill waste plus olive leaves (OL; ratio 1:1 *w/w*) were used for the cultivation of *Pleurotus* strains. The WS substrate was provided by Dirfis Mushrooms IKE (Kathenoi, Euboea, Greece), grape marc was obtained from a winery in the Nemea area (northeast Peloponnese, Greece), and the two-phase olive-mill wastes and olive leaves from an olive-oil mill situated in Kalamata (southwest Peloponnese, Greece), respectively. Substrates were milled to a particle size of 2–3 cm and soaked in water for 24 h. Water surplus was drained off (moisture content of the substrates was 53%–69%), and the substrates were mixed with calcium carbonate and wheat bran (2% *w/w* and 5% *w/w*, respectively). Two kg of each formulated substrate was then placed into autoclavable polypropylene bags and sterilized twice for 1 h (121 °C, 1.1 atm). Inoculation of substrates was performed with a spawn (5% *w/w*) prepared as described by Koutrotsios et al. [3]. Four replicates per substrate were used. Incubation of cultures and fructification were carried out in a specially-designed mushroom cultivation room under conditions previously reported [3]. Prior to the analyses, the collected mushroom samples were freeze-dried and grinded to a particle size less than 2 mm.

### 2.3. Determination of Glucan and Ergosterol Content

The determination of the mushrooms' total and α-glucans content was performed by the Mushroom and Yeast Beta-Glucan assay kit (Megazyme Int., Bray, Ireland) according to the manufacturer's instructions, while the β-glucans content was calculated by subtracting α-glucans from the total glucans. Light absorbance was measured at 510 nm using a Hitachi U-2001 spectrophotometer (Hitachi High-Tech America, Inc.; Schaumburg, IL, USA).

The mushrooms' ergosterol content was determined as described by Sapozhnikova et al. [34]. Cholesterol (100 µg/mL, internal standard) was added in 100–200 mg of the freeze-dried mushroom sample and saponified with 2 mL of potassium hydroxide (3M) in methanol under sonication (10 min) and heating (60 °C, 60 min). All manipulations were performed under reduced light conditions to avoid the potential conversion of ergosterol to vitamin $D_2$. The un-saponified fraction was extracted twice with 3 mL of hexane. Hexane extracts were then pooled and evaporated to dryness (Speed Vac, Labconco Corporation, Kansas City, MO, USA). Sterols were derivatized to trimethylsilylethers (TMS) with N,O-Bis(trimethylsilyl)trifluoroacetamide (BSTFA) at 70 °C for 20 min, and 1 µL aliquots were injected in the gas chromatographer (Agilent HP GC 6890 N; Wallborn, Germany) coupled with a mass spectrometer (Agilent HP 5973; Wallborn, Germany) at a split ratio of 5:1. The analysis of the TMS sterol derivatives was carried out under electron impact ionization (70 eV) and separated by an Agilent J&W HP-5ms capillary column (30 m × 0.25 mm × 250 µm) with a carrier gas flow rate equal to 0.6 mL/min (high-purity He). The injector and MS detector transfer line were kept at 220 °C and 300 °C. The oven temperature was set initially at 210 °C, raised to 300 °C at 5.5 °C/min, and held for 14 min. The identity of ergosterol was verified by the presence of expected ion fragments at the proper ratios according to literature [35,36]. Ergosterol quantification was performed by constructing a 6-point calibration curve, covering the range 0–600 µg, and by employing cholesterol as an internal standard.

### 2.4. Attenuated Total Reflection—Fourier Transform Infrared (ATR-FTIR) Analysis

ATR-FTIR spectra of the mushroom samples were obtained by a Perkin Elmer Spectrum-Two spectrometer equipped with a Diamond ATR compartment (Perkin Elmer, Hopkinton, MA, USA) using the Spectrum 10 software provided by the manufacturer. For each sample, 32 scans of the infrared region between 4000 and 400 $cm^{-1}$ at a resolution of 4 $cm^{-1}$ were recorded in triplicates and averaged. The recorded spectra were then ATR-corrected with a refractive index for diamond of 1.5 in order to be comparable to the available spectral libraries for facilitating the interpretation of spectra.

A spectroscopic analysis followed to obtain a comparative insight among the mushrooms produced by the different *Pleurotus* species on various substrates. Prior to the spectroscopic analysis, the spectra were smoothed by the Savitzky–Golay algorithm (5 points each side (total window of 11 smoothing points) and a zero order polynomial) [37], linear baseline corrected and then normalized by the mean using The Unscrambler X v.10.5 software (CAMO software, Oslo, Norway).

## 2.5. Multivariate Analysis

A principal component analysis (PCA) on the smoothed, baseline-corrected and normalized ATR-FTIR spectra of the mushrooms was performed to detect any grouping in terms of species or cultivation substrate by using The Unscrambler X v.10.5 software (CAMO software, Oslo, Norway). For this purpose, singular value decomposition (SVD) was applied for 20 principal components using a leave-one-out cross-validation. Partial least square regression (PLSR) analysis was performed to calibrate models predicting the glucans and ergosterol contents of mushrooms on the basis of their recorded ATR-FTIR spectra. A wide range of spectral transformations and various combinations were applied to the recorded spectra (i.e., Savitzky–Golay smoothing, smoothing by the median, linear and non-linear baseline correction, normalization by the mean, multiplicative scatter correction, standard normal variate, de-trending, first and second derivative, etc.) to obtain better predictions. Potential sample outliers were detected using the interquartile ranges approach [38] for the measured glucans and ergosterol values, while spectral outliers were identified by Hotelling's $T^2$ distribution [39]. In order to avoid overestimations in predictions, the sample sets for each model (i.e., 79 samples for the glucans prediction and 109 samples for the ergosterol prediction) were divided into a calibration (CAL) set containing nine tenths of the samples and an external validation (EV) set with the remaining samples. The CAL set was used to develop the calibration model on which the optimal number of components was chosen based on a leave-one-out cross-validation (CV; models' self-testing). The EV set was constructed by selecting every tenth sample following the order of the glucans or ergosterol contents in the mushrooms, and used for the evaluation of the robustness of the developed models. Non-significant variables were removed in some cases by the Martens' uncertainty test [40] to improve the models' stability and robustness. The Unscrambler X v.10.5 software (CAMO software, Oslo, Norway) was used for all calibrations.

The models' performance was determined by the $R^2$ (coefficient of determination) value (Equation (1)):

$$R^2 = \frac{\Sigma_i(y_i - f_i)^2}{\Sigma_i(y_i - \bar{y})^2} \qquad (1)$$

where $y_i$ represents the measured values and $f_i$ represents the predicted values. The closer $R^2$ is to 1, the better the fit of the measured values ($y_i$) to the regression line.

The models' precision was determined by the root mean square error (RMSE) in % of the dry weight (dw) for the glucans content and mg g$^{-1}$ (dw) for the ergosterol content (Equation (2)):

$$RMSE = \sqrt{\frac{\Sigma_{i=0}^{n}(f_i - y_i)^2}{n}} \qquad (2)$$

where $y_i$ represents the measured values and $f_i$ represents the predicted values.

In addition, the relative error of prediction (REP) given in % [41] was calculated by (Equation (3))

$$REP = 100\frac{RMSE}{z} \qquad (3)$$

where $z$ is the mean value of the calibration concentrations for the analyte examined.

## 3. Results and Discussion

### 3.1. Glucan and Ergosterol Contents of Pleurotus Species

The total glucan content of the *P. ostreatus* strains ranged from 38.84% to 58.90% of dry weight (dw) for the mushrooms produced on the WS substrate, from 28.28% to 48.42% dw for the GM substrate, and from 15.53% to 41.16% dw for the OL substrate. The values range for the β-glucan contents (total glucans minus α-glucans) was 30.18–48.16% dw (WS), 22.66–40.56% dw (GM) and 14.62–31.31% dw (OL). As regards the ergosterol content, the values range measured for the *P. ostreatus* mushrooms was 6.42–16.06 mg g$^{-1}$ dw (WS), 10.94–26.09 mg g$^{-1}$ dw (GM) and 11.82–20.25 mg g$^{-1}$ dw (OL). For *P. eryngii*, the respective contents varied less than in *P. ostreatus*, i.e., the total and β-glucans contents for the mushrooms deriving from the WS substrate were 32.84–61.40% dw and 26.44–51.36% dw, respectively, from the GM substrate were 37.08–54.11% dw and 32.18–44.73% dw, respectively, while from the OL substrate they were 31.02–52.69% dw and 27.54–42.33% dw, respectively. A similar pattern was also observed for the *P. eryngii* strains in respect to the mushrooms' ergosterol content, which was 4.83–14.26 mg g$^{-1}$ dw (WS), 7.30–14.10 mg g$^{-1}$ dw (GM) and 9.27–19.42 mg g$^{-1}$ dw (OL). As regards the *P. nebrodensis* mushrooms grown on WS and GM, the total glucans were 38.72% and 44.86% dw, while the β-glucans were 30.23% and 35.11% dw, respectively. The ergosterol content was 13.91 mg g$^{-1}$ dw and 12.43 mg g$^{-1}$ dw in the *P. nebrodensis* mushrooms from WS and GM, respectively.

The mean of the measured total, α-, β-glucan and ergosterol contents of the *Pleurotus* strains was projected for *P. ostreatus* and *P. eryngii* (Figure 1) in order to obtain a generalized perspective of the effect that different cultivation substrates have on the content of bioactive compounds. The *P. ostreatus* strains revealed a significant decrease in terms of the total and β-glucan contents when substrates other than WS were used (i.e., WS > GM > OL), whereas a significant increase was observed for the ergosterol content in GM and OL (Figure 1a). A similar pattern was observed for the *P. eryngii* strains as regards both glucans and ergosterol (Figure 1b). However, no significant differences between the cultivation substrates were observed for β-glucans and ergosterol as it was the case in *P. ostreatus*, which might be indicative of a reduced impact that the cultivation media could exert on the *P. eryngii* mushroom content in these compounds.

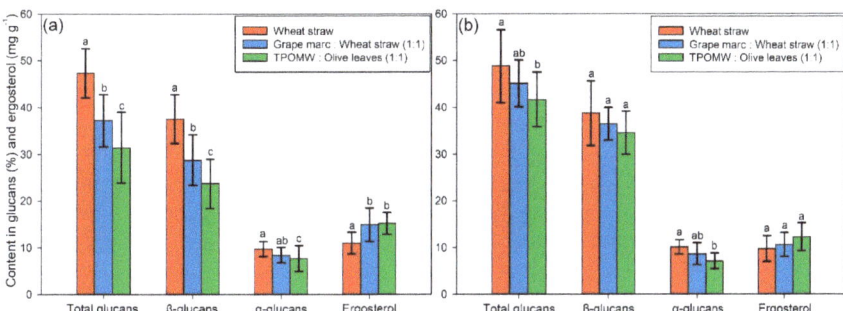

**Figure 1.** Mushroom contents in the total, α- and β-glucans (% dw), and ergosterol (mg g$^{-1}$ dw) for (**a**) *Pleurotus ostreatus* and (**b**) *P. eryngii* cultivated in three substrates, i.e., wheat straw (control; WS), grape marc plus wheat straw (1:1 *w/w*; GM) and two-phase olive mill waste (TPOMW) plus olive leaves (1:1 *w/w*; OL). Error bars represent standard deviation among the strains of each species. Lack of letters in common indicates statistically significant differences (Duncan's *t*-Test, $p < 0.05$) in comparisons of bioactive compounds content between different substrates for each species examined.

## 3.2. Qualitative Analysis of Pleurotus Mushrooms Based on ATR-FTIR Spectroscopy

### 3.2.1. Spectral Comparison of *P. ostreatus* and *P. eryngii* Mushrooms Cultivated on Different Substrates

In order to perform a comparative evaluation among the mushrooms produced on different substrates, the recorded spectra of the *P. ostreatus* strains on each substrate were averaged (Figure 2a). Differences were observed in the IR absorption regions at 3316, 1641, 1548, 1400 and 1200–1050 cm$^{-1}$. Similar peaks were also detected for the *P. eryngii* mushrooms, while an additional peak was evident at 1745 cm$^{-1}$ (Figure 2b). The peak at 3316 cm$^{-1}$ (observed in mushrooms produced in WS, and shifted to 3313 and 3301 cm$^{-1}$ for GM and OL, respectively) can be attributed to the N-H stretching vibration of the amide A band in proteins and nucleic acids or the O-H stretching vibration in phenols and $H_2O$ [42,43]. The IR absorption peak at 1745 cm$^{-1}$, which was only observed in the spectra of the *P. eryngii* mushrooms, corresponds to the C=O stretching of phospholipids. A similar pattern (i.e., presence in *P. eryngii*, absence in *P. ostreatus*) was also reported by Zervakis et al. [29].

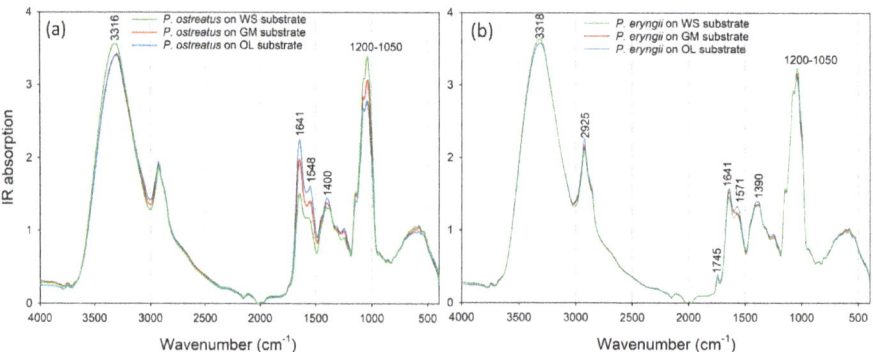

**Figure 2.** Recorded attenuated total reflectance (ATR)-FTIR spectra of (**a**) *Pleurotus ostreatus* and (**b**) *P. eryngii* mushrooms cultivated on wheat straw (WS), wheat straw plus grape marc mix, 1:1 *w/w* (GM) and two-phase olive mill waste plus olive leaves, 1:1 *w/w* (OL) substrates.

The peak at 1641 cm$^{-1}$ could be associated with the C=O stretching vibration in the amide I band [42], the C=C and C=O stretching vibrations in amino acids [42,43], the N-H bending in flavonoids [43] and the aromatic ring deformations [44]. This specific peak was found to be well correlated with the antioxidant capacity of the propolis samples [45]; therefore, the higher absorption intensity for mushrooms produced on OL and GM is indicative of a higher antioxidant activity of the *Pleurotus* mushrooms deriving from these particular substrates, as previously evidenced [13]. Additionally, the higher absorption intensity in this region for mushrooms cultivated on OL, followed by those originating from GM and WS, is in agreement with their total phenolic content and concurs with previous pertinent findings [13]. The same also applies for the *P. eryngii* mushrooms; however, smaller differences were detected in the IR absorption signal for this peak since the effect of the cultivation substrate on the total phenolic content is not as pronounced as in *P. ostreatus*. The peak around 1550 cm$^{-1}$ could be assigned to the N-H bending vibration and C-N stretching vibration of the amide II region in proteins, while the peak around 1400 cm$^{-1}$ can be associated with the symmetric stretching vibration of the COO- group of fatty acids and amino acids, the symmetric bending modes of methyl groups in skeletal proteins and the symmetric stretch of methyl groups in proteins [42]. The last two peaks (i.e., at 1550 cm$^{-1}$ and 1400 cm$^{-1}$) could indicate a higher protein content in the mushrooms produced on the OL and GM substrates.

Koutrotsios et al. [3] reported an increase in the total crude protein content of the *P. ostreatus* mushrooms cultivated on the OL substrate. The region at 1200–900 cm$^{-1}$ could be assigned to the C-O stretching vibration of the pyranose compounds in carbohydrates [42,44], with the absorption in this region to be more intense for the mushrooms produced on WS (followed by GM and OL), indicating a higher content in polysaccharides. This is in agreement with the glucans content measured in the respective samples (Figure 1a). The differences in this particular region are rather low in the *P. eryngii* mushrooms; however, they are in accordance with their glucans content (Figure 1b), as in the case of *P. ostreatus*.

### 3.2.2. Comparative Evaluation of Pleurotus Species

A spectroscopic comparison of the *P. ostreatus*, *P. eryngii* and *P. nebrodensis* mushrooms cultivated on WS was performed to detect potential spectroscopic differences among the species examined (Figure 3).

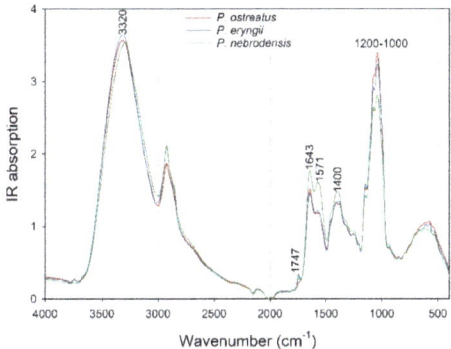

**Figure 3.** Recorded ATR-FTIR spectra for the *Pleurotus ostreatus*, *P. eryngii* and *P. nebrodensis* mushrooms cultivated on wheat straw (WS).

Indeed, differences were revealed in the spectral regions at 3320, 1747, 1643, 1571, 1400 and 1200–1000 cm$^{-1}$, which were particularly obvious in the case of *P. nebrodensis*. The peak at 1747 cm$^{-1}$ was evident only in the *P. eryngii* and *P. nebrodensis* spectra, and could be attributed to the C=O stretching of phospholipids. The fact that this peak was produced by these two species only (and not by *P. ostreatus*) is in agreement to their close phylogenetic affinity [46]. The peaks at 1643 cm$^{-1}$ (C=O stretching in the amide I region, C=C and C=O stretching in amino acids, N-H bending in flavonoids and aromatic ring deformations), 1571 cm$^{-1}$ (N-H bending and C-N stretching in the amide II region in proteins) and 1400 cm$^{-1}$ (COO- group symmetric stretching in fatty acids and amino acids, methyl groups symmetric bending in skeletal proteins and methyl group symmetric stretch in proteins) had an increased IR absorption for *P. nebrodensis*, which could indicate an increased protein content of this particular strain in respect to the *P. ostreatus* and *P. eryngii* material. Finally, the region at 1200–1000 cm$^{-1}$, which is related to the C-O stretching vibration of the pyranose compounds in carbohydrates, revealed an increasing IR absorption intensity from *P. nebrodensis* to *P. eryngii* to *P. ostreatus*, in accordance with the total glucans content determined for these particular species.

### 3.3. Principal Component Analysis (PCA)

A PCA was performed on the ATR-FTIR spectra of the *P. ostreatus*, *P. eryngii* and *P. nebrodensis* mushrooms to detect groupings/associations of interest. Most of the spectral variance (>99%) was explained through the first ten principal components (PCs), with the first three explaining 91% of the variance (i.e., PC1: 67.4%; PC2: 13.2%; PC3: 10.4%). The correlation of the first (PC1) and third principal components (PC3) revealed a fairly clear separation of the *Pleurotus* species on the basis of the recorded spectra (Figure 4a), which is mostly evident across the *y*-axis (PC3).

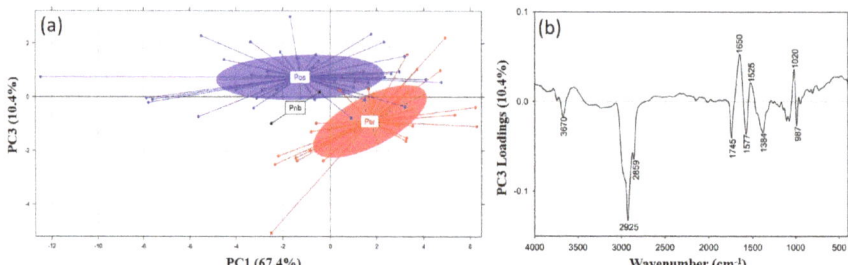

**Figure 4.** (**a**) Score plot of principal component analysis (PCA) (PC1 vs. PC3) for the discrimination of the *Pleurotus* species (Pos: *P. ostreatus*; Per: *P. eryngii*; Pnb: *P. nebrodensis*) on the basis of their recorded ATR-FTIR spectra, and (**b**) PCA loadings for the PC3 across which discrimination is evident.

The PC3 loadings (Figure 4b) were interpreted to identify the spectral regions responsible for this separation, and subsequently to identify potentially related compounds. The *P. eryngii* and *P. nebrodensis* mushrooms exhibited similar spectroscopic characteristics and positively correlated with regions at 2925 and 2859 cm$^{-1}$ (aliphatic compounds), 1745 cm$^{-1}$ (phospholipids), 1577 cm$^{-1}$ (amide II region) and 1384 cm$^{-1}$ (C-N stretching in tertiary aromatic amines; CH$_3$ symmetric vibrations in lipids).

On the other hand, the *P. ostreatus* mushrooms were positively correlated with regions at 1650 cm$^{-1}$ (N-H bending of primary amines and C=O stretching in the amide I region) [42,44,47], 1525 cm$^{-1}$ (amide II region) and 1020 cm$^{-1}$ (pyranose compounds in carbohydrates). The latter region is related to the polysaccharide content of mushrooms and is indicative of the higher glucans content in *P. ostreatus*, while the region at 1745 cm$^{-1}$, which was positively correlated to *P. eryngii* and *P. nebrodensis*, was also found to characterize these two species (Figure 3).

Furthermore, a PCA was performed for discriminating the mushrooms on the basis of the substrate on which they were produced. Even if discrimination was not so clear, as it was among species, probably due to the similar characteristic compounds present in the mushrooms, the different IR absorption intensities allowed a partial separation, especially of the mushrooms produced on WS and OL through the correlation of the first two PCs (PC1 vs. PC2) (Figure 5a). An interpretation of the PC1 loadings (Figure 5b), across which discrimination was mostly evident, revealed a dominant positive correlation (i.e., the group positioned to the right, positive numbers of this axis) with the region 1200–950 cm$^{-1}$ (polysaccharide region). This fact, combined with the previously reported higher content in glucans for the mushrooms cultivated on WS (Figure 1), confirms the accuracy of this separation.

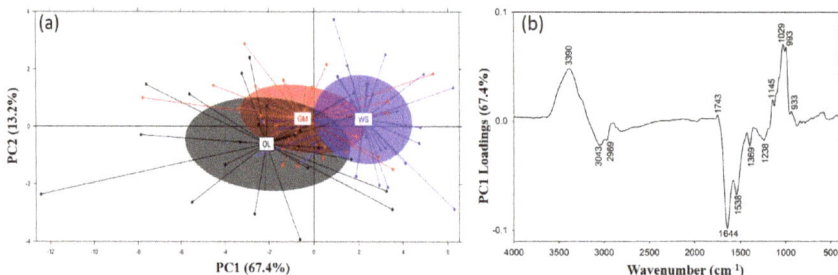

**Figure 5.** (**a**) Score plot of PCA (PC1 vs. PC2) for the discrimination of substrates (WS: wheat straw; GM: grape marc plus wheat straw, 1:1 *w/w*; OL: two-phase olive mill waste plus olive leaves, 1:1 *w/w*) on which the *Pleurotus* mushrooms were produced on the basis of their recorded ATR-FTIR spectra, and (**b**) PCA loadings for the PC1 across of which the discrimination is evident.

## 3.4. Prediction of Mushrooms Glucans Content

PLSR models were developed for the prediction of the total and β-glucans contents of the mushrooms on the basis of their ATR-FTIR spectra. Among various spectral transformations performed prior to the calibration, Savitzky–Golay smoothing (window of 11 smoothing points, zero polynomial) combined with lineal baseline correction and normalization by the mean provided the best precision and accuracy of the calibrated model. The latter, predicting the total glucans content in the *Pleurotus* mushrooms, was developed from a set of 72 samples, leaving out seven samples to be used for the external validation of the model's performance, accuracy and robustness. A sample was identified as an outlier based on the interquartile ranges approach as well as on the observation of the score plot of the reference vs. the predicted values [48]. For six factors, the developed model achieved an $R^2$ value of 0.90 and 0.84 for the calibration ($R^2_{CAL}$) and cross-validation ($R^2_{CV}$), respectively, with root mean square error values of 2.77% ($RMSE_{CAL}$) and 3.44% ($RMSE_{CV}$) (Figure 6a). The respective values for the external validation sample set were 0.83 ($R^2_{EV}$) and 2.46% ($RMSE_{EV}$). Based on that, the relative error of prediction for the external validation ($REP_{EV}$) was calculated at 5.8%, which has been previously characterized as a highly acceptable error value for calibrated models [49].

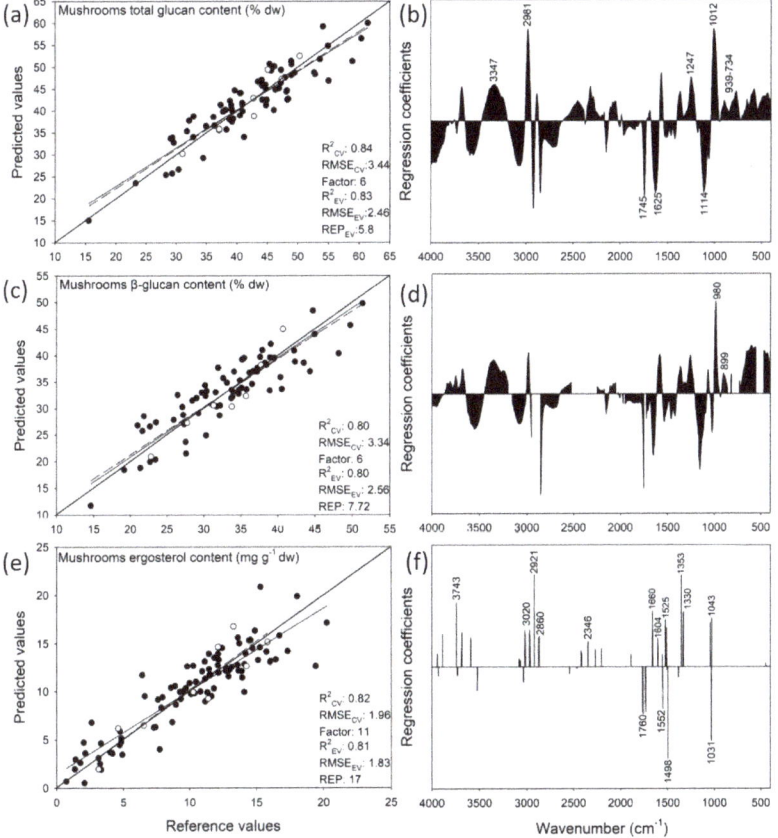

**Figure 6.** The measured vs. the predicted values and regression coefficients of calibration for the prediction model of (**a**,**b**) the total glucans content, (**c**,**d**) the β-glucans content and (**e**,**f**) the ergosterol content in the *Pleurotus* mushrooms.

The developed model predicting the β-glucan content (Figure 6b) was also calibrated for 72 samples (seven samples left out to be used for the external validation), while the same sample set was also identified as an outlier on the basis of the score plot of the reference vs. the predicted values and Hotelling's $T^2$ distribution [39,48]. A removal of non-significant variables was applied using the jack-knifing algorithm [40]. For six factors, the developed model achieved an $R^2$ value of 0.86 for the calibration ($R^2_{CAL}$) and 0.80 for cross-validation ($R^2_{CV}$), while the root mean square values were 2.73% and 3.34%, respectively. For the external validation sample set, the achieved $R^2_{EV}$ value was 0.80 with a $RMSE_{EV}$ of 2.56%. Furthermore, the calculated relative error of prediction for the external validation predictions was 7.72%, pointing to a reasonable/acceptable error for this calibrated model.

In order to increase the model's robustness and regression coefficients (i.e., the spectral regions automatically selected by The Unscrambler software for the calibrations and correlated positively or negatively with the developed prediction models) were interpreted, as a way of eliminating as much as possible the possibility of an artifact (i.e., model calibrated based on irrelevance to the predicted value spectral regions). The interpretation of the regression coefficients revealed that the prediction of the mushrooms' total glucans content (Figure 6b) was significantly positively correlated with the spectral regions at 3347 cm$^{-1}$ (OH symmetric and asymmetric stretching), 2981 cm$^{-1}$ (methyl group C-H stretching), 1247 cm$^{-1}$ (among others C-H stretching and O-H deformations in carbohydrates), 1012 cm$^{-1}$ (C-O stretching in carbohydrates) and 939–734 cm$^{-1}$ (C-H vibrations related to the α- and β-pyranose compounds, both glycosidic and non-glycosidic) [44]. The former region (i.e., at 3347 cm$^{-1}$) may also refer to the OH groups which are contained at a significant number in the backbone of glucans [50].

On the other hand, a significant negative correlation of the predicted total glucans content could be observed at the regions of 1745 cm$^{-1}$ (C=O in phospholipids), 1625 cm$^{-1}$ (C=O stretching in the amide I band) and 1114 cm$^{-1}$ (C-O stretching in crystalline cellulose) [42,51]. The interpretation of the regression coefficients used for the prediction of the β-glucans (Figure 6d) revealed very similar correlations with the regression coefficients for the total glucans predictions, but in this case the polysaccharide region (between 1200 and 800 cm$^{-1}$) has been replaced by a very strong positive correlation at 980 cm$^{-1}$ and by a positive correlation at 899 cm$^{-1}$. Socrates [44] assigned these regions to the symmetric ring vibration and the C-H deformation, and designated them as characteristic of the β-pyranose compounds.

Unfortunately, due to lack of studies related to the prediction of the total and β-glucans in the mushrooms on the basis of the IR spectroscopic data, a direct comparison of the model developed in this study was not feasible. Ma et al. [52] used near infrared spectroscopy (NIRS) to predict the polysaccharide content in mycelia of *Ganoderma* species and achieved an $R^2_{CV}$ value of 0.98. Nevertheless, this model was developed for fungi phylogenetically well-separated from *Pleurotus*; more importantly, it was based on the outcome of the analysis of mycelia (and not of fruitbodies) for which the compositional properties can be different [53]. In addition, a different spectroscopic technique was adopted (i.e., NIRS), which reflects mid-infrared overtone regions and combination bands that can be highly overlapping [54]. In this way, very limited information is obtained in respect to the chemical components associated with the regions used for the calibration; hence, a direct identification of the compounds is difficult [55]. Finally, the prediction of the total polysaccharides content makes the model less specific since it includes a wide range of sugar compounds. In a relevant study, Chen et al. [53] reported $R^2$ values of 0.973–0.989 and an $RMSE_P$ = 0.225–0.012 for a model developed on the basis of NIR spectra of polysaccharides and triterpenoids in *Ganoderma lucidum* and *G. atrum* mushrooms. When the plant material was examined, Gracia et al. [56] obtained a high prediction performance ($R^2$ values) of the β-glucan content using NIRS for a sample set of 1728 single intact groats of six different oat varieties, while Brown et al. [57] predicted the total glucans content in *Setaria viridis* plants by using ATR-FTIR with $R^2$ values of 0.90 for a sample set of 183 collections. In addition, Li et al. [58] used FTIR for the prediction of the total polysaccharides content in Chinese ginseng (*Panax notoginseng*), achieving an $R^2$ value of 0.83 for the external validation sample set with relatively a low REP. Although the outcome of these studies is not directly comparable to the model developed in the present work, since different materials were examined, they are indicative of its good predictive power. It is noteworthy that as far as mushrooms are concerned, FTIR and NIR spectra have been also used to build prediction models suitable for addressing other issues, e.g., the post-harvest quality deterioration of *A. bisporus* fruitbodies [33,59] and the geographical traceability of *Boletus* spp. [60,61].

### 3.5. Prediction of Mushrooms Ergosterol Content

The PLSR models predicting the mushrooms' content in ergosterol were developed on the previously transformed spectra by detrend (second polynomial) and first Savitzky–Golay derivation (window of 11 smoothing points and second polynomial). The calibrated model predicting the ergosterol content in the *Pleurotus* mushrooms was developed on a set of 100 samples, leaving out nine samples for the external validation of the model's performance, accuracy and robustness. Ten samples from the calibration sample set were identified as the outliers on the basis of the interquartile ranges approach as well as on observations of the score plot of the reference vs. the predicted values and the Hotelling's $T^2$ distribution [39,48]. A removal of non-significant variables was also applied by using the jack-knifing algorithm of Martens and Martens [40], provided by The Unscrambler software. For 11 factors, the developed model achieved an $R^2_{CAL}$ value of 0.90 and an $R^2_{CV}$ of 0.82, while the root mean square values were 1.47 mg g$^{-1}$ dw and 1.96 mg g$^{-1}$ dw, respectively. For the external validation sample set, the achieved $R^2_{EV}$ and $RMSE_{EV}$ values were 0.81 and 1.83 mg g$^{-1}$ dw, respectively (Figure 6e). However, the calculated relative error of prediction for the external validation prediction was approximately 17%, which makes the enhancement of the model's predictive accuracy necessary, even if it reveals a relative reasonable error. This can be potentially achieved by increasing the number of samples included during calibration.

The interpretation of the regression coefficients used during calibration (Figure 6f) would significantly improve the trustfulness of this model, if related ergosterol spectral regions could be used. To address this issue, the spectrum of ergosterol standard compound (Sigma-Aldrich, EC Number 200-352-7, CAS Number 57-87-4) was recorded by using the same spectroscopic setup, and peaks characteristic of ergosterol were detected (Figure 7). As it was found, the negatively correlated regression coefficients used for the prediction of ergosterol were not observed as characteristic peaks in the ergosterol spectrum, while most of the positively correlated regression coefficients (i.e., at 3020, 2921, 2860, 1660, 1353, 1330 and 1043 cm$^{-1}$) corresponded to peaks of the ergosterol spectrum (Figure 7). This revealed a potentially high validity/robustness of the model developed for the prediction of ergosterol as it derived from the spectral regions of the respective standard compound. A direct comparison with previous studies was again not possible since to the best of the authors' knowledge, it is the first time that a model predicting fungal ergosterol content on the basis of acquired spectroscopic data has been presented. Recently, Shapaval et al. [62] reported a prediction of the total lipid content in oleaginous yeasts by applying high-throughput FTIR spectroscopy and achieved a high $R^2$ value for the cross-validation (0.92); however, the $R^2$ value achieved during the external validation was 0.67. Moreover, the total sterol content in brown algae was estimated with fairly good accuracy but through the use of a single peak of the FTIR spectrum in conjunction with a previously calibrated standard regression line [63].

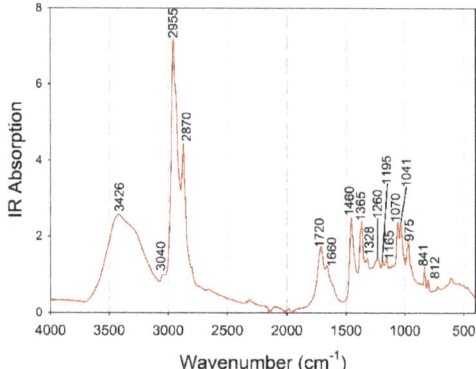

**Figure 7.** ATR-FTIR spectrum of the ergosterol standard with marked characteristic peaks.

## 4. Conclusions

The outcome of the present study indicates that ATR-FTIR could serve as a potential PAT tool for the estimation of glucans and ergosterol contents in *Pleurotus* mushrooms; therefore, it can be exploited as an alternative to the laborious, costly and/or time-consuming analytical techniques/assays used until now. The glucans models could be identified as accurate (low RMSE, low REP and representative regression coefficients) and of a high performance (good $R^2$), while the RMSE and REP of the ergosterol models can be further improved with the addition of more samples (even though the regression coefficients used in this work for the calibration were extremely accurate). In general, PAT processes and prediction models are dynamic and need to be constantly updated/fed with new entries to sustain/improve their performance. In addition, ATR-FTIR successfully characterized the mushroom samples by detecting differences related to the species or substrates used, and by separating them into groups through a principal component analysis.

**Author Contributions:** Conceptualization, G.B. and G.I.Z.; mushroom cultivation, G.K.; sample preparation, G.K. and G.B.; glucan measurement, G.B.; determination of ergosterol content, D.T. and N.K.; acquisition of ATR-FTIR spectra, G.B.; developments of prediction models, G.B.; writing—original draft preparation, G.B.; writing—review and editing, G.K., D.T., N.K. and G.I.Z.; visualization, G.B.; supervision, G.I.Z. and N.K.; project administration, G.I.Z.; funding acquisition, G.I.Z. and N.K. All authors have read and agreed to the published version of the manuscript.

**Funding:** This research has been co-financed by the European Union and Greek national funds (European Social Fund—SF) through the Operational Program Competitiveness, Entrepreneurship and Innovation, under the call RESEARCH–CREATE–INNOVATE (project code: T1EDK–02560).

**Acknowledgments:** Substrates used as controls were kindly donated by Dirfis Mushrooms IKE (Evvoia).

**Conflicts of Interest:** The authors declare no conflict of interest.

## References

1. Royse, D.J.; Baars, J.; Tan, Q. Current overview of mushroom production in the world. In *Edible and Medicinal Mushrooms*; Diego, C.Z., Pardo-Giménez, A., Eds.; Wiley Online Library location: Hoboken, NJ, USA, 2017; pp. 5–13.
2. Koutrotsios, G.; Larou, E.; Mountzouris, K.C.; Zervakis, G.I. Detoxification of olive mill wastewater and bioconversion of olive crop residues into high-value-added biomass by the choice edible mushroom *Hericium erinaceus*. *Biotechnol. Appl. Biochem.* **2016**, *180*, 195–209. [CrossRef]
3. Koutrotsios, G.; Mountzouris, K.C.; Chatzipavlidis, I.; Zervakis, G.I. Bioconversion of lignocellulosic residues by *Agrocybe cylindracea* and *Pleurotus ostreatus* mushroom fungi–Assessment of their effect on the final product and spent substrate properties. *Food Chem.* **2014**, *161*, 127–135. [CrossRef] [PubMed]

4. Julian, A.V.; Reyes, R.G.; Eguchi, F. Agro-industrial waste conversion into medicinal mushroom cultivation. In *Encyclopedia of Environmental Health*, 2nd ed.; Nriagu, J., Ed.; Elsevier: Oxford, UK, 2019; pp. 13–20.
5. Zervakis, G.I.; Koutrotsios, G. Solid-state fermentation of plant residues and agro-industrial wastes for the production of medicinal mushrooms. In *Medicinal Plants and Fungi: Recent Advances in Research and Development*; Agrawal, D.C., Tsay, H.-S., Shyur, L.-F., Wu, Y.-C., Wang, S.-Y., Eds.; Springer: Singapore, 2017; pp. 365–396.
6. Mediterranean Plant Specialist Group. The Top 50 Mediterranean Island Plants UPDATE 2017. Available online: http://top50.iucn-mpsg.org/species/39 (accessed on 23 March 2020).
7. Gargano, M.L.; van Griensven, L.J.L.D.; Isikhuemhen, O.S.; Lindequist, U.; Venturella, G.; Wasser, S.P.; Zervakis, G.I. Medicinal mushrooms: Valuable biological resources of high exploitation potential. *Plant Biosyst. Int. J. Deal. All Asp. Plant Biol.* 2017, *151*, 548–565. [CrossRef]
8. Synytsya, A.; Míčková, K.; Synytsya, A.; Jablonský, I.; Spěváček, J.; Erban, V.; Kováříková, E.; Čopíková, J. Glucans from fruit bodies of cultivated mushrooms *Pleurotus ostreatus* and *Pleurotus eryngii*: Structure and potential prebiotic activity. *Carbohydr. Polym.* 2009, *76*, 548–556. [CrossRef]
9. Sari, M.; Prange, A.; Lelley, J.I.; Hambitzer, R. Screening of beta-glucan contents in commercially cultivated and wild growing mushrooms. *Food Chem.* 2017, *216*, 45–51. [CrossRef] [PubMed]
10. Koutrotsios, G.; Patsou, M.; Mitsou, E.K.; Bekiaris, G.; Kotsou, M.; Tarantilis, P.A.; Pletsa, V.; Kyriacou, A.; Zervakis, G.I. Valorization of olive by-products as substrates for the cultivation of *Ganoderma lucidum* and *Pleurotus ostreatus* mushrooms with enhanced functional and prebiotic properties. *Catalysts* 2019, *9*, 537. [CrossRef]
11. De Silva, D.D.; Rapior, S.; Fons, F.; Bahkali, A.H.; Hyde, K.D. Medicinal mushrooms in supportive cancer therapies: An approach to anti-cancer effects and putative mechanisms of action. *Fungal Divers.* 2012, *55*, 1–35. [CrossRef]
12. Jasinghe, V.J.; Perera, C.O.; Sablani, S.S. Kinetics of the conversion of ergosterol in edible mushrooms. *J. Food Eng.* 2007, *79*, 864–869. [CrossRef]
13. Koutrotsios, G.; Kalogeropoulos, N.; Kaliora, A.C.; Zervakis, G.I. Toward an increased functionality in oyster (*Pleurotus*) mushrooms produced on grape marc or olive mill wastes serving as sources of bioactive compounds. *J. Agric. Food Chem.* 2018, *66*, 5971–5983. [CrossRef]
14. Koutrotsios, G.; Kalogeropoulos, N.; Stathopoulos, P.; Kaliora, A.C.; Zervakis, G.I. Bioactive compounds and antioxidant activity exhibit high intraspecific variability in *Pleurotus ostreatus* mushrooms and correlate well with cultivation performance parameters. *World J. Microbiol. Biotechnol.* 2017, *33*, 98. [CrossRef]
15. Hoa, H.T.; Wang, C.L.; Wang, C.H. The effects of different substrates on the growth, yield, and nutritional composition of two oyster mushrooms (*Pleurotus ostreatus* and *Pleurotus cystidiosus*). *Mycobiology* 2015, *43*, 423–434. [CrossRef] [PubMed]
16. Ntougias, S.; Gaitis, F.; Katsaris, P.; Skoulika, S.; Iliopoulos, N.; Zervakis, G.I. The effects of olives harvest period and production year on olive mill wastewater properties—Evaluation of *Pleurotus* strains as bioindicators of the effluent's toxicity. *Chemosphere* 2013, *92*, 399–405. [CrossRef] [PubMed]
17. Tournour, H.H.; Segundo, M.A.; Magalhães, L.M.; Barreiros, L.; Queiroz, J.; Cunha, L.M. Valorization of grape pomace: Extraction of bioactive phenolics with antioxidant properties. *Ind. Crop. Prod.* 2015, *74*, 397–406. [CrossRef]
18. Romero, C.; Medina, E.; Mateo, M.A.; Brenes, M. New by-products rich in bioactive substances from the olive oil mill processing. *J. Sci. Food Agric.* 2018, *98*, 225–230. [CrossRef]
19. Teixeira, A.; Baenas, N.; Dominguez-Perles, R.; Barros, A.; Rosa, E.; Moreno, D.A.; Garcia-Viguera, C. Natural bioactive compounds from winery by-products as health promoters: A review. *Int. J. Mol. Sci.* 2014, *15*, 15638–15678. [CrossRef]
20. Cullen, P.J.; O'Donnell, C.P.; Fagan, C.C. Benefits and challenges of adopting PAT for the food industry. In *Process Analytical Technology for the Food Industry*; O'Donnell, C.P., Fagan, C., Cullen, P.J., Eds.; Springer: New York, NY, USA, 2014; pp. 1–5.
21. Roussel, S.; Preys, S.; Chauchard, F.; Lallemand, J. Multivariate data analysis (chemometrics). In *Process Analytical Technology for the Food Industry*; O'Donnell, C.P., Fagan, C., Cullen, P.J., Eds.; Springer: New York, NY, USA, 2014; pp. 7–59.

22. Venetsanou, A.; Anastasaki, E.; Gardeli, C.; Tarantilis, P.A.; Pappas, C.S. Estimation of antioxidant activity of different mixed herbal infusions using attenuated total reflectance Fourier transform infrared spectroscopy and chemometrics. *Emir. J. Food Agric.* **2017**, *29*, 149–155. [CrossRef]
23. Sousa, N.; Moreira, M.J.; Saraiva, C.; De Almeida, J.M.M.M. Applying Fourier transform mid infrared spectroscopy to detect the adulteration of *Salmo salar* with *Oncorhynchus mykiss*. *Foods* **2018**, *7*, 55. [CrossRef]
24. Rodriguez-Saona, L.E.; Allendorf, M.E. Use of FTIR for rapid authentication and detection of adulteration of food. *Annu. Rev. Food Sci. Technol.* **2011**, *2*, 467–483. [CrossRef]
25. Erwanto, Y.; Muttaqien, A.T.; Sugiyono; Sismindari; Rohman, A. Use of Fourier transform infrared (FTIR) spectroscopy and chemometrics for analysis of lard adulteration in "Rambak" crackers. *Int. J. Food Prop.* **2016**, *19*, 2718–2725. [CrossRef]
26. Fagan, C.C. Infrared spectroscopy. In *Process Analytical Technology for the Food Industry*; O'Donnell, C.P., Fagan, C., Cullen, P.J., Eds.; Springer: New York, NY, USA, 2014; pp. 73–101.
27. Lecellier, A.; Mounier, J.; Gaydou, V.; Castrec, L.; Barbier, G.; Ablain, W.; Manfait, M.; Toubas, D.; Sockalingum, G.D. Differentiation and identification of filamentous fungi by high-throughput FTIR spectroscopic analysis of mycelia. *Int. J. Food Microbiol.* **2014**, *168*, 32–41. [CrossRef]
28. Santos, C.; Fraga, M.E.; Kozakiewicz, Z.; Lima, N. Fourier transform infrared as a powerful technique for the identification and characterization of filamentous fungi and yeasts. *Res. Microbiol.* **2010**, *161*, 168–175. [CrossRef] [PubMed]
29. Zervakis, G.I.; Bekiaris, G.; Tarantilis, P.; Pappas, C.S. Rapid strain classification and taxa delimitation within the edible mushroom genus *Pleurotus* through the use of diffuse reflectance infrared Fourier transform (DRIFT) spectroscopy. *Fungal Biol.* **2012**, *116*, 715–728. [CrossRef] [PubMed]
30. Wang, Y.Y.; Li, J.Q.; Liu, H.G.; Wang, Y.Z. Attenuated total reflection-Fourier transform infrared spectroscopy (ATR-FTIR) combined with chemometrics methods for the classification of Lingzhi species. *Molecules* **2019**, *24*, 2210. [CrossRef] [PubMed]
31. Yao, S.; Li, J.; Li, T.; Liu, H.; Wang, Y. Discrimination of Boletaceae mushrooms based on data fusion of FT-IR and ICP–AES combined with SVM. *Int. J. Food Prop.* **2018**, *21*, 255–266. [CrossRef]
32. Chen, Y.; Xie, M.Y.; Yan, Y.; Zhu, S.B.; Nie, S.P.; Li, C.; Wang, Y.X.; Gong, X.F. Discrimination of *Ganoderma lucidum* according to geographical origin with near infrared diffuse reflectance spectroscopy and pattern recognition techniques. *Anal. Chim. Acta* **2008**, *618*, 121–130. [CrossRef]
33. O'Gorman, A.; Downey, G.; Gowen, A.A.; Barry-Ryan, C.; Frias, J.M. Use of Fourier transform infrared spectroscopy and chemometric data analysis to evaluate damage and age in mushrooms (*Agaricus bisporus*) grown in Ireland. *J. Agric. Food Chem.* **2010**, *58*, 7770–7776. [CrossRef]
34. Sapozhnikova, Y.; Byrdwell, W.C.; Lobato, A.; Romig, B. Effects of UV-B radiation levels on concentrations of phytosterols, ergothioneine and polyphenolic compounds in mushroom powders used as dietary supplements. *J. Agric. Food Chem.* **2014**, *62*, 3034–3042. [CrossRef]
35. Phillips, K.M.; Ruggio, D.M.; Horst, R.L.; Minor, B.; Simon, R.R.; Feeney, M.J.; Byrdwell, W.C.; Haytowitz, D.B. Vitamin D and sterol composition of 10 types of mushrooms from retail suppliers in the United States. *J. Agric. Food Chem.* **2011**, *59*, 7841–7853. [CrossRef]
36. Teichmann, A.; Dutta, P.C.; Staffas, A.; Jägerstad, M. Sterol and vitamin $D_2$ concentrations in cultivated and wild grown mushrooms: Effects of UV irradiation. *Lwt Food Sci. Technol.* **2007**, *40*, 815–822. [CrossRef]
37. Savitzky, A.; Golay, M.J.E. Smoothing and differentiation of data by simplified least squares procedures. *Anal. Chem.* **1964**, *36*, 1627–1639. [CrossRef]
38. Vinutha, H.P.; Poornima, B.; Sagar, B.M. *Detection of Outliers Using Interquartile Range Technique from Intrusion Dataset*; Springer: Singapore, 2018; pp. 511–518.
39. Hotelling, H. The generalization of Student's ratio. *Inst. Math. Stat.* **1931**, 360–378. [CrossRef]
40. Martens, H.; Martens, M. Modified Jack-knife estimation of parameter uncertainty in bilinear modelling by partial least squares regression (PLSR). *Food Qual. Prefer.* **2000**, *11*, 5–16. [CrossRef]
41. Olivieri, A.C. The classical least-squares model. In *Introduction to Multivariate Calibration: A Practical Approach*; Springer International Publishing: Cham, Germany, 2018; pp. 19–38.
42. Movasaghi, Z.; Rehman, S.; Rehman, I.U. Fourier transform infrared (FTIR) spectroscopy of biological tissues. *Appl. Spectrosc. Rev.* **2008**, *43*, 134–179. [CrossRef]

43. Oliveira, R.N.; Mancini, M.C.; Oliveira, F.C.S.D.; Passos, T.M.; Quilty, B.; Thiré, R.M.D.S.M.; McGuinness, G.B. FTIR analysis and quantification of phenols and flavonoids of five commercially available plants extracts used in wound healing. *Matéria* **2016**, *21*, 767–779. [CrossRef]
44. Socrates, G. *Infrared and Raman Characteristic Group Frequencies: Tables and Charts*, 3rd ed.; John Wily & Sons Ltd.: Chichester, UK, 2001.
45. Moţ, A.C.; Silaghi-Dumitrescu, R.; Sârbu, C. Rapid and effective evaluation of the antioxidant capacity of propolis extracts using DPPH bleaching kinetic profiles, FT-IR and UV–vis spectroscopic data. *J. Food Compos. Anal.* **2011**, *24*, 516–522. [CrossRef]
46. Zervakis, G.I.; Ntougias, S.; Gargano, M.L.; Besi, M.I.; Polemis, E.; Typas, M.A.; Venturella, G. A reappraisal of the *Pleurotus eryngii* complex–New species and taxonomic combinations based on the application of a polyphasic approach, and an identification key to *Pleurotus* taxa associated with *Apiaceae* plants. *Fungal Biol.* **2014**, *118*, 814–834. [CrossRef]
47. Coates, J. Interpretation of infrared spectra: A practical approach interpretation of infrared spectra. In *Encyclopedia of Analytical Chemistry*; Meyers, R.A., Ed.; John Wiley & Sons, Ltd.: Hoboken, NJ, USA, 2000; pp. 10815–10837.
48. Bro, R.; Rinnan, Å.; Faber, N.M. Standard error of prediction for multilinear PLS: 2. Practical implementation in fluorescence spectroscopy. *Chemom. Intell. Lab. Syst.* **2005**, *75*, 69–76. [CrossRef]
49. Olivieri, A.C. Chemometrics and multivariate calibration. In *Introduction to Multivariate Calibration: A Practical Approach*; Springer International Publishing: Cham, Germanty, 2018; pp. 1–17.
50. Fazio, A.; La Torre, C.; Caroleo, M.C.; Caputo, P.; Plastina, P.; Cione, E. Isolation and purification of glucans from an italian cultivar of *Ziziphus jujuba* Mill. and in vitro effect on skin repair. *Molecules* **2020**, *25*, 968. [CrossRef]
51. Ciolacu, D.; Ciolacu, F.; Popa, V.I. Amorphous cellulose-structure and characterization. *Cellul. Chem. Technol.* **2011**, *45*, 13–21.
52. Ma, Y.; He, H.; Wu, J.; Wang, C.; Chao, K.; Huang, Q. Assessment of polysaccharides from mycelia of genus *Ganoderma* by mid-infrared and near-infrared spectroscopy. *Sci. Rep.* **2018**, *8*, 10. [CrossRef]
53. Chen, Y.; Xie, M.; Zhang, H.; Wang, Y.; Nie, S.; Li, C. Quantification of total polysaccharides and triterpenoids in *Ganoderma lucidum* and *Ganoderma atrum* by near infrared spectroscopy and chemometrics. *Food Chem.* **2012**, *135*, 268–275. [CrossRef]
54. Engelsen, S.B. Near infrared spectroscopy—A unique window of opportunities. *Nir. News* **2016**, *27*, 14–17. [CrossRef]
55. Ríos-Reina, R.; García-González, D.L.; Callejón, R.M.; Amigo, J.M. NIR spectroscopy and chemometrics for the typification of Spanish wine vinegars with a protected designation of origin. *Food Control* **2018**, *89*, 108–116. [CrossRef]
56. Gracia, M.-B.; Armstrong, P.R.; Rongkui, H.; Mark, S. Quantification of betaglucans, lipid and protein contents in whole oat groats (*Avena sativa* L.) using near infrared reflectance spectroscopy. *J. Near Infrared Spectrosc.* **2017**, *25*, 172–179. [CrossRef]
57. Brown, C.; Martin, A.P.; Grof, C.P.L. The application of Fourier transform mid-infrared (FTIR) spectroscopy to identify variation in cell wall composition of *Setaria italica* ecotypes. *J. Integr. Agric.* **2017**, *16*, 1256–1267. [CrossRef]
58. Li, Y.; Zhang, J.; Liu, F.; Xu, F.; Wang, Y.; Zhang, J.-Y. Prediction of total polysaccharides content in *P. notoginseng* using FTIR combined with SVR. *Spectrosc. Spectr. Anal.* **2018**, *38*, 1696–1701. [CrossRef]
59. Esquerre, C.; Gowen, A.A.; O'Donnell, C.P.; Downey, G. Initial studies on the quantitation of bruise damage and freshness in mushrooms using visible-near-infrared spectroscopy. *J. Agric. Food Chem.* **2009**, *57*, 1903–1907. [CrossRef] [PubMed]
60. Li, Y.; Zhang, J.; Li, T.; Liu, H.; Li, J.; Wang, Y. Geographical traceability of wild *Boletus edulis* based on data fusion of FT-MIR and ICP-AES coupled with data mining methods (SVM). *Spectrochim. Acta Part A Mol. Biomol. Spectrosc.* **2017**, *177*, 20–27. [CrossRef]
61. Li, Y.; Wang, Y. Synergistic strategy for the geographical traceability of wild *Boletus tomentipes* by means of data fusion analysis. *Microchem. J.* **2018**, *140*, 38–46. [CrossRef]

62. Shapaval, V.; Brandenburg, J.; Blomqvist, J.; Tafintseva, V.; Passoth, V.; Sandgren, M.; Kohler, A. Biochemical profiling, prediction of total lipid content and fatty acid profile in oleaginous yeasts by FTIR spectroscopy. *Biotechnol. Biofuels* **2019**, *12*, 140. [CrossRef]
63. Bouzidi, N.; Daghbouche, Y.; El Hattab, M.; Aliche, Z.; Culioli, G.; Piovetti, L.; Garrigues, S.; de la Guardia, M. Determination of total sterols in brown algae by Fourier transform infrared spectroscopy. *Anal. Chim. Acta* **2008**, *616*, 185–189. [CrossRef] [PubMed]

© 2020 by the authors. Licensee MDPI, Basel, Switzerland. This article is an open access article distributed under the terms and conditions of the Creative Commons Attribution (CC BY) license (http://creativecommons.org/licenses/by/4.0/).

*Article*

# Fast Quantification of Honey Adulteration with Laser-Induced Breakdown Spectroscopy and Chemometric Methods

Jiyu Peng [1], Weiyue Xie [1], Jiandong Jiang [1], Zhangfeng Zhao [1], Fei Zhou [2,4,*] and Fei Liu [3]

[1] Key Laboratory of E & M (Zhejiang University of Technology), Ministry of Education & Zhejiang Province, Hangzhou 310014, China; jypeng@zjut.edu.cn (J.P.); wyxiee@163.com (W.X.); jiangjd@zjut.edu.cn (J.J.); i12fly@163.com (Z.Z.)
[2] College of Standardization, China Jiliang University, Hangzhou 310018, China
[3] College of Biosystems Engineering and Food Science, Zhejiang University, Hangzhou 310058, China; fliu@zju.edu.cn
[4] Beingmate (Hangzhou) Food Research Institute Co., Ltd, Hangzhou 311106, China
* Correspondence: feizhou@cjlu.edu.cn; Tel.: +86-15-068-8290-30

Received: 17 February 2020; Accepted: 11 March 2020; Published: 14 March 2020

**Abstract:** Honey adulteration is a major issue in food production, which may reduce the effective components in honey and have a detrimental effect on human health. Herein, laser-induced breakdown spectroscopy (LIBS) combined with chemometric methods was used to fast quantify the adulterant content. Two common types of adulteration, including mixing acacia honey with high fructose corn syrup (HFCS) and rape honey, were quantified with univariate analysis and partial least squares regression (PLSR). In addition, the variable importance was tested with univariable analysis and feature selection methods (genetic algorithm (GA), variable importance in projection (VIP), selectivity ratio (SR)). The results indicated that emissions from Mg II 279.58, 280.30 nm, Mg I 285.25 nm, Ca II 393.37, 396.89 nm, Ca I 422.70 nm, Na I 589.03, 589.64 nm, and K I 766.57, 769.97 nm had compact relationship with adulterant content. Best models for detecting the adulteration ratio of HFCS 55, HFCS 90, and rape honey were achieved by SR-PLSR, VIP-PLSR, and VIP-PLSR, with root-mean-square error (RMSE) of 8.9%, 8.2%, and 4.8%, respectively. This study provided a fast and simple approach for detecting honey adulteration.

**Keywords:** honey; adulteration; feature variable; partial least square regression; laser-induced breakdown spectroscopy

## 1. Introduction

Food adulteration is an illegal activity of food production, which may threaten food quality and safety. On one hand, the nutritional value of food is limited because of the reduction of effective components in food. On the other hand, the adulterants may have a detrimental effect on human health. Several scandals concerning food adulteration have been reported around the world [1–3]. Honey is one of the most commonly adulterated foods because of its economical purpose and wide use. There are two main approaches for honey adulteration. One is to mix pure honey with sugar-based adulterants, and the other is to adulterate high-quality honey with inferior honey. These two cases will be explored in this study.

The adulterant usually has a similar constituent or characteristic with the pure honey, and it is hard to distinguish from the appearance. Several studies concerning honey adulteration detection have been reported. Amiry et al. [4] discriminated adulterated honey (mix pure honey with date syrup and invert sugar syrup) with linear discriminant analysis. Different parameters including color indices, rheological, physical, and chemical parameters were used as variables for discrimination. Physical and chemical parameters achieved the best results, with accuracy above 95%. The results highlighted the use of physical and chemical parameters to detect honey adulteration. In addition, Arroyo-Manzanares et al. [5] used gas chromatography-ion mobility spectrometry to detect sugar cane or corn syrup adulterated honey; seven out of nine commercial honeys were classified as adulterated samples. Traditionally, the chemical features of honey are detected with wet chemical analysis, which is time and labor consuming. Hence, several rapid analytical methods based on electronic and optical techniques were proposed by other researchers, e.g., electronic nose [6], electronic tongue [6], fluorescence spectroscopy [7], visible-near infrared spectroscopy [8,9]. The 'fingerprint information' of honey could be rapidly obtained by these sensors, and the adulterated honey could be distinguished with the help of chemometric methods.

For its part, laser-induced breakdown spectroscopy (LIBS), which allows elemental analysis, may be useful for honey authenticity. The elemental information of honey can be obtained through analyzing the atomic emission spectroscopy from plasma which is induced by a laser. It has the advantages of fast detection, multi-elemental analysis, and environmentally friendly feature [10]. As a novel approach in food, it has been used for regional discrimination [11] and elemental detection [12–14]. Because LIBS spectrum often contains numerous variables, chemometric methods are usually used to figure out the useful information and establish models for food adulteration detection. Recently, LIBS was used to classify the botanical origins of honey, and detect rice syrup adulterated samples [15]. However, the adulterant content in honey should be further quantified. Herein, LIBS combined with partial least squares regression was used as an analytical tool for fast quantification of honey adulterant content.

In this study, acacia honey mixed with high fructose corn syrup (HFCS) and rape honey were analyzed by LIBS. The specific objectives were to: (1) analyze the LIBS spectral features of pure honey and adulterants; (2) determine the feature variables that are related to adulteration; (3) quantify the adulterant content with univariate and multivariate analysis.

## 2. Materials and Methods

### 2.1. Sample Preparation

Honey including acacia honey (Guanshengyuan Co., Ltd, Shanghai, China) and rape honey (Yaoquan Food Co., Ltd, Yunnan, China) were collected from main producers in China, and two kinds of HFCS with different fructose concentrations (F55 and F90) were purchased from markets. HFCS F55 contains 55% fructose, and HFCS F90 contains 90% fructose. In this case, acacia honey was considered as pure honey, and HFCS (F55 and F90) and rape honey were used as adulterants.

Honey adulteration was prepared by mixing the acacia honey with HFCS F55, HFCS F90, and rape honey. To establish models for quantifying adulterant content, acacia honey was adulterated with HFCS and rape honey at 21 different percentages (0%, 5%, 10%, 15%, 20%, 25%, 30%, 35,% 40%, 45%, 50%, 55%, 60%, 65%, 70%, 75%, 80%, 85%, 90%, 95%, and 100%). In addition, adulterated samples for external prediction were prepared at 13 different adulteration rates, i.e., 0%, 8%, 16%, 24%, 32%, 40%, 48%, 56%, 64%, 72%, 80%, 88%, and 96%. The adulteration rates of 0% and 100% indicated pure acacia honey and pure adulterant, respectively. All sample adulteration was performed in three replications, so there were 63 samples for calibration, and 39 samples for prediction. After mixing, all samples were kept in a water bath at 37 °C for 12 h to ensure homogeneity.

### 2.2. LIBS Measurement

A laboratory-assembled LIBS device was used for honey adulteration detection. The detailed description of the device was introduced in our previous published article [16]. First, 8 g of sample was added in 12-well plates and placed in a X-Y-Z moving stage. A pulse laser (Vlite 200, Beamtech, Beijing, China) operated at 532 nm was used to ablate the sample with energy of 80 mJ. Then, emission light from induced plasma was transferred into an Echelle spectrograph (ME 5000, Andor, Belfast, UK), and detected by an intensified charge coupled device (ICCD, DH334T-18F-03, Andor, Belfast, UK). To improve the signal-to-background ratio, the delay time, integral time, and relative gain of ICCD camera were set at 2 µs, 10 µs, and 26. Single shot scanning was performed in an ablation region of 10 mm × 10 mm with resolution of 1 mm. Hence, 100 successive spectra were collected for each sample, the spectra were averaged to minimize the sample inhomogeneity. Because of the advantages of LIBS, no sample preparation was needed, and the total detection time for one sample was less than two minutes.

### 2.3. Data Analysis

Because the peak in LIBS spectrum corresponds to the emission from a certain element or molecule band, the observed peak intensity was used as the variable for analysis. To establish a model for quantifying adulterant content, PLSR was used. In addition, several feature selection methods based on PLSR were used to determine the key LIBS emissions that related to the adulterant content.

PLSR is a commonly and widely used multivariate method for quantitative analysis. It projects the raw variables into new dimensions with the maximal variation, and regresses the first few new variables (latent variable, LV) with respond value [17]. In this case, the raw variables were peak intensities of main emissions, and the respond value was the adulterant content in honey. Before modeling, the auto scale preprocessing method, which used mean-centering followed by dividing each variable by the standard variation of the variable, was used to correct the scaling of each variable. Ten-folds random cross-validation was used to determine the number of LV, and prevent the overfitting. In addition, the straightforward implementation of a statistically inspired modification of the PLS (SIMPLS) algorithm was used to calculate the PLS model parameters [18].

Three feature selection methods including genetic algorithm (GA), variable importance in projection (VIP), and the selectivity ratio (SR) were used in this case. GA is a subset search algorithm that was inspired by biological evolution theory and natural selection [19]. The subset of relevant variables selected by GA is then fitted with PLSR to evaluate the performance, and determine the feature variables. Different from GA, the variable selection based on VIP and SR is carried out by using a threshold of some parameters from the PLSR model. VIP calculates the accumulation of PLS weights, and SR defines the ratio between explained variance and the unexplained variance in the PLS model. The larger values of VIP and SR, the greater contribution of the variable. For the criteria of variable selection, VIP follows the rule of 'greater than one rule', and SR follows the F-test (95%) criterion [20]. In this case, the variables with VIP value greater than 1 and SR value greater than 1.532 were selected as important variables.

After modeling, some measures should be used to evaluate the performance. In this case, model performance was evaluated with correlation coefficient ($r$) and root-mean-square error (RMSE). The $r$ value measures the relationship between predicted adulterant content and actual value, and the RMSE value measures the predictive error. The larger the $r$ value and the smaller the RMSE value, the better the model performance. All data analyses were carried out in the MATLAB (v2019b, The MathWorks Inc., Natick, MA, USA).

## 3. Results and Discussion

### 3.1. LIBS Spectral Characteristics

Before quantification, LIBS spectral characteristics of acacia honey, rape honey, HFCS F55, and HFCS F90 were first analyzed (Figure 1). All the LIBS spectra ranged from 240 to 860 nm. In general, the average LIBS spectra for different samples were similar except some emissions in certain spectral range. It was credited to the similar constituent of honey and HFCS. In general, honey contains 75% saccharides (mainly glucose and fructose), 15% water, amino acids, and minerals, etc. HFCS mainly contains glucose and fructose. According to the concentration of fructose, the HFCS can be divided into three categories: F42 (42% fructose), F55 (55% fructose), and F90 (90% fructose). Hence, the main components ablated by laser in both honey and HFCS were glucose and fructose. As shown in Figure 1, the emissions from C, H, O, and N were observed in all samples. The molecular band CN that usually appears in an organic sample when analyzed in air atmosphere was also found in this case.

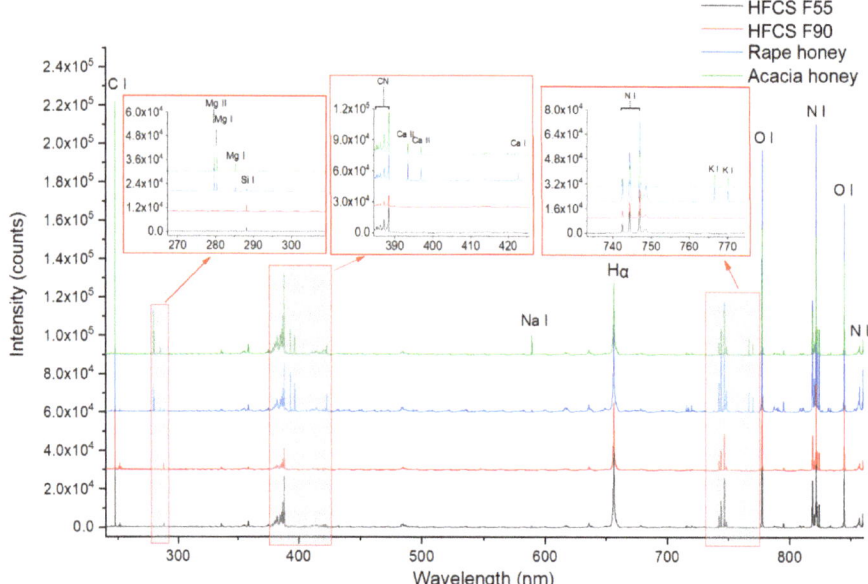

**Figure 1.** Average laser-induced breakdown spectroscopy (LIBS) spectrum of honey (acacia honey and rape honey) and high fructose corn syrup (HFCS 55 and HFCS 90).

Some differences in elemental emissions could be observed between honey and HFCS. It was obvious that emissions from Mg, Ca, and K appeared in the spectra of honey, while it cannot be found in the spectra of HFCS. It indicated that the concentrations of Mg, Ca, and K in honey were significantly higher than those in HFCS. In addition, there was no obvious difference between acacia honey and rape honey, except relatively stronger emission of Na in acacia honey. These elemental differences might be used to differentiate the adulterants. However, it was hard to quantify the adulterant content simply by analyzing spectrum. Hence, some modeling methods were further used to quantify the adulterant content.

### 3.2. Univariate Analysis

Univariate analysis was used to explore the relationship between adulterant content and single variable and quantify the adulteration. In this case, the peak intensities of main emissions from samples were used for analysis. Univariate analysis was performed by regressing the peak intensity of each emission with the adulterant content, and $r$ and RMSE were used to evaluate the results. The corresponding element for each emission could be identified with the National Institute of Standard and Technology (NIST, Gaithersburg, Maryland, USA) database [21]. Table 1 shows the results of univariate analysis between main emission lines and adulterant content. Forty-three univariate models were established. The variables contained emissions from C, Si, Mg, Ca, Na, K, N, H, O, and CN. Four variables with emissions of 748.47, 794.83, 795.17, and 822.43 nm were marked with unknown, because they could not be identified with the NIST database or references.

In general, the models for quantifying adulterant content of HFCS F90 had the best results with higher $r$ and lower RMSE. It indicated that high concentration of fructose in HFCS led to greater spectral difference and contributed to the univariate analysis. In addition, for HFCS F90 and HFCS F55, the emissions from Mg II 279.58, 280.30 nm, Mg I 285.25 nm, Ca II 393.37, 396.89 nm, Ca I 422.70 nm,

Na I 589.03, 589.63 nm, and K I 766.57, 769.97 nm had compact relationship with the adulterant content, with $r > 0.9$ and RMSE < 11.0%. For rape honey, models based on emissions from Na I 589.03 and 589.63 nm had good results, with $r$ of 0.919 and 0.903, and RMSE of 12.0% and 13.0%. It indicated that emissions from mineral elements played an important role in adulteration quantification. It also verified the LIBS spectral difference between acacia honey and adulterants.

Table 1. Results of univariate analysis based on peak intensities of main emissions.

| No. | Observed Wavelength (nm) | Element | HFCS F55 | | HFCS F90 | | Rape Honey | |
|---|---|---|---|---|---|---|---|---|
| | | | r | RMSE | r | RMSE | r | RMSE |
| 1 | 247.88 | C I | 0.493 | 26.4% | 0.823 | 17.2% | 0.066 | 30.2% |
| 2 | 250.72 | Si I | 0.176 | 29.8% | 0.204 | 29.6% | 0.154 | 29.9% |
| 3 | 251.45 | Si I | 0.180 | 29.8% | 0.214 | 29.6% | 0.153 | 29.9% |
| 4 | 251.64 | Si I | 0.204 | 29.7% | 0.222 | 29.5% | 0.166 | 29.8% |
| 5 | 251.94 | Si I | 0.193 | 29.7% | 0.205 | 29.6% | 0.159 | 29.9% |
| 6 | 252.44 | Si I | 0.195 | 29.7% | 0.210 | 29.6% | 0.149 | 29.9% |
| 7 | 252.88 | Si I | 0.200 | 29.7% | 0.210 | 29.6% | 0.158 | 29.9% |
| 8 | 279.58 | Mg II | 0.932 | 10.9% | 0.936 | 10.6% | 0.516 | 25.9% |
| 9 | 280.30 | Mg II | 0.922 | 11.7% | 0.934 | 10.8% | 0.441 | 27.2% |
| 10 | 285.25 | Mg I | 0.959 | 8.6% | 0.959 | 8.6% | 0.517 | 25.9% |
| 11 | 288.20 | Si I | 0.194 | 29.7% | 0.227 | 29.5% | 0.161 | 29.9% |
| 12 | 385.07 | CN 4-4 | 0.550 | 25.3% | 0.828 | 17.0% | 0.487 | 26.4% |
| 13 | 385.49 | CN 3-3 | 0.576 | 24.8% | 0.820 | 17.3% | 0.454 | 27.0% |
| 14 | 386.17 | CN 2-2 | 0.596 | 24.3% | 0.821 | 17.3% | 0.442 | 27.2% |
| 15 | 387.13 | CN 1-1 | 0.473 | 26.7% | 0.828 | 17.0% | 0.460 | 26.9% |
| 16 | 388.33 | CN 0-0 | 0.514 | 26.0% | 0.824 | 17.2% | 0.466 | 26.8% |
| 17 | 393.37 | Ca II | 0.957 | 8.8% | 0.948 | 9.6% | 0.694 | 21.8% |
| 18 | 396.89 | Ca II | 0.959 | 8.6% | 0.951 | 9.3% | 0.652 | 22.9% |
| 19 | 422.70 | Ca I | 0.953 | 9.2% | 0.942 | 10.1% | 0.707 | 21.4% |
| 20 | 589.03 | Na I | 0.937 | 10.6% | 0.973 | 6.9% | 0.919 | 12.0% |
| 21 | 589.64 | Na I | 0.936 | 10.6% | 0.975 | 6.8% | 0.903 | 13.0% |
| 22 | 656.33 | Hα | 0.617 | 23.8% | 0.538 | 25.5% | 0.243 | 29.4% |
| 23 | 715.77 | O I | 0.316 | 28.7% | 0.766 | 19.5% | 0.227 | 29.5% |
| 24 | 742.45 | N I | 0.220 | 29.5% | 0.739 | 20.4% | 0.268 | 29.2% |
| 25 | 744.30 | N I | 0.197 | 29.7% | 0.738 | 20.4% | 0.278 | 29.1% |
| 26 | 746.92 | N I | 0.162 | 29.9% | 0.742 | 20.3% | 0.248 | 29.3% |
| 27 | 748.47 | Unknown | 0.507 | 26.1% | 0.632 | 23.4% | 0.220 | 29.5% |
| 28 | 766.57 | K I | 0.943 | 10.1% | 0.960 | 8.4% | 0.756 | 19.8% |
| 29 | 769.97 | K I | 0.931 | 11.1% | 0.959 | 8.6% | 0.750 | 20.0% |
| 30 | 777.47 | O I | 0.183 | 29.8% | 0.760 | 19.7% | 0.215 | 29.6% |
| 31 | 794.83 | Unknown | 0.316 | 28.7% | 0.758 | 19.7% | 0.215 | 29.6% |
| 32 | 795.17 | Unknown | 0.299 | 28.9% | 0.773 | 19.2% | 0.206 | 29.6% |
| 33 | 818.57 | N I | 0.170 | 29.9% | 0.740 | 20.4% | 0.256 | 29.3% |
| 34 | 818.86 | N I | 0.217 | 29.6% | 0.736 | 20.5% | 0.265 | 29.2% |
| 35 | 820.10 | N I | 0.232 | 29.5% | 0.746 | 20.2% | 0.237 | 29.4% |
| 36 | 821.14 | N I | 0.221 | 29.5% | 0.744 | 20.2% | 0.247 | 29.3% |
| 37 | 821.68 | N I | 0.244 | 29.4% | 0.725 | 20.8% | 0.250 | 29.3% |
| 38 | 822.28 | N I | 0.067 | 30.2% | 0.782 | 18.9% | 0.305 | 28.8% |
| 39 | 822.43 | Unknown | 0.290 | 29.0% | 0.706 | 21.4% | 0.303 | 28.8% |
| 40 | 824.32 | N I | 0.291 | 29.0% | 0.719 | 21.0% | 0.285 | 29.0% |
| 41 | 844.73 | O I | 0.252 | 29.3% | 0.743 | 20.3% | 0.260 | 29.2% |
| 42 | 856.86 | N I | 0.325 | 28.7% | 0.729 | 20.7% | 0.275 | 29.1% |
| 43 | 859.49 | N I | 0.357 | 28.3% | 0.706 | 21.5% | 0.316 | 28.7% |

Note: The shade color of the table represents the performance of univariate analysis. The shade color of being green indicates the best compact relationship ($r = \pm 1$) and the lowest predictive error (RMSE = 0).

### 3.3. Quantification of Adulterant Content Based on Multivariate Analysis

Multivariate analysis was further used to quantify the adulterant content. First, all variables in univariate analysis were used as the inputs of PLS models. As seen in Table 2, PLS models based on all variables achieved good results for all three types of adulteration. The $r$ values for HFCS F55, HFCS

F90, rape honey in the prediction set were 0.962, 0.980, 0.988, and the RSME values were 15.6%, 16.6%, 4.7%, respectively. The latent variables for these three models were 4, 4, 5, which were determined by cross validation. The results of PLS models were better than those of univariate analysis. It also verified the advantages of multivariate analysis. The combination of information from multiple emissions contributed to the adulterant content quantification.

Table 2. Multivariate analysis results based on partial least square regression (PLSR) and feature selection methods.

| Adulterant | Method | No. of LV | No. of Var. | Calibration | | C.V. | | Prediction | |
|---|---|---|---|---|---|---|---|---|---|
| | | | | r | RMSE | r | RMSE | r | RMSE |
| HFCS F55 | PLSR | 4 | 43 | 0.977 | 6.5% | 0.965 | 8.0% | 0.962 | 15.6% |
| | GA-PLSR | 4 | 12 | 0.983 | 5.6% | 0.978 | 6.4% | 0.794 | 32.0% |
| | VIP-PLSR | 5 | 16 | 0.982 | 5.7% | 0.966 | 8.1% | 0.938 | 18.6% |
| | **SR-PLSR** | **1** | **11** | **0.965** | **7.9%** | **0.960** | **8.5%** | **0.966** | **8.9%** |
| HFCS F90 | PLSR | 4 | 43 | 0.973 | 7.0% | 0.964 | 8.2% | 0.980 | 16.6% |
| | GA-PLSR | 5 | 19 | 0.979 | 6.1% | 0.972 | 7.3% | 0.985 | 11.3% |
| | **VIP-PLSR** | **5** | **15** | **0.982** | **5.7%** | **0.977** | **6.5%** | **0.980** | **8.2%** |
| | SR-PLSR | 5 | 20 | 0.981 | 5.9% | 0.973 | 7.0% | 0.982 | 9.4% |
| Rape honey | PLSR | 5 | 43 | 0.993 | 3.6% | 0.990 | 4.3% | 0.988 | 4.7% |
| | GA-PLSR | 4 | 21 | 0.994 | 3.3% | 0.990 | 4.4% | 0.988 | 4.7% |
| | **VIP-PLSR** | **3** | **10** | **0.991** | **4.1%** | **0.989** | **4.6%** | **0.988** | **4.8%** |
| | SR-PLSR | 1 | 2 | 0.912 | 12.4% | 0.874 | 15.0% | 0.943 | 11.3% |

Note: No. of LV: number of latent variables; No. of var.: number of variables; C.V.: cross-validation; r: correlation coefficient; RMSE: root-mean-square error; GA: genetic algorithm; VIP: variable importance in projection; SR: selectivity ratio.

In addition, results of PLS models based on feature variables (selected by GA, VIP, and SR) are also shown in Table 2. In general, prediction results after feature selection were similar or better than those based all variables. The irrelevant variables in models might worsen the modeling performance [22,23], which also verified the necessity of feature selection. Only one exception happened for the GA-PLS model in HFCS F55 quantification. The RMSE value in prediction set was 0.320, which is greatly worse than that without feature selection (0.156). It might be credited to the selected variables by the GA method. As shown in Figure 2, lots of irrelevant variables were selected. The GA method might not be suitable for feature selection in the honey adulteration with HFCS F55. With the consideration of variable number and prediction performance, the models marked with bold achieved the best results. The RMSE value for HFCS 55, HFCS F90, and rape honey in the prediction set were 8.9%, 8.2%, and 4.8%, respectively. In addition, similar results were achieved in 10-folds cross-validation, and RMSE value for HFCS 55, HFCS F90, and rape honey were 8.5%, 6.5%, and 4.6%, respectively.

We also compared the variables selected with GA, VIP, and SR methods (Figure 2). Row 1, 5, 9 showed the correlation coefficient between each variable and adulterant content of HFCS F55, HFCS F90, and rape honey, respectively. The values of correlation coefficient were in the range of 0 to 1. Other rows represented the variables selected by GA, VIP, and SR methods. Selected variables were represented in blue, and non-selected variables were in white. As shown in Figure 2, VIP and SR methods chose the variables with a high correlation coefficient, while some variables with a low correlation coefficient were selected by the GA method. It was related to the principal of feature selection methods. For the GA method, the variables were randomly combined and verified by PLSR. The variables were selected based on the results of PLSR modeling. For VIP and SR methods, the contribution of each variable was considered in the selection [20]. The variables selected by the GA method might be easily affected when testing with external samples. In addition, VIP and SR methods had some common variables, while the number of selected variables was different. It might be credited to the different threshold measure of each method. Hence, VIP and SR methods might be recommended for feature selection in quantification of honey adulterant content.

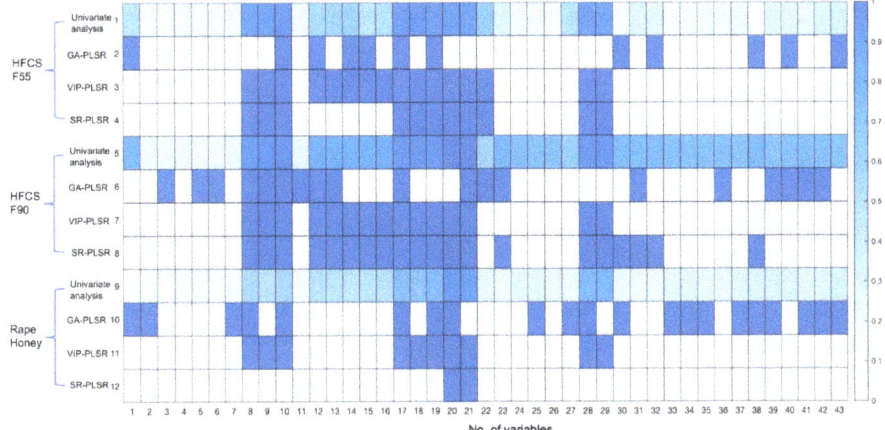

**Figure 2.** Feature variables selected with genetic algorithm (GA), variable importance in projection (VIP), and selectivity ratio (SR) methods. Row 1, 5, 9 shows the univariate analysis result between each variable and adulterant content of HFCS F55, HFCS F90, and rape honey, respectively. Cells with a gradient of blue color indicated the correlation coefficient. Other rows represented the variables selected by GA, VIP, and SR methods. Selected variables were represented in blue, and non-selected variables were in white.

The scatter plot of the best model for quantifying adulteration ratio of HFCS 55, HFCS 90, and rape honey is shown in Figure 3. Among these three models, the quantification for rape honey achieved the best result, with $r$ and RMSE of 0.988 and 4.8% in the prediction set. The samples in calibration and prediction sets distributed closely around the regression lines, and the regression lines almost went through original point. The emissions from Mg II 279.58, 280.30 nm, Mg I 285.25 nm, Ca II 393.37, 396.89 nm, Ca I 422.70 nm, Na I 589.03, 589.64 nm, and K I 766.57, 769.97 nm, which were the feature variables in the rape honey quantification, were also included in the other two models. It indicated that these variables might play an important role in honey adulteration analysis.

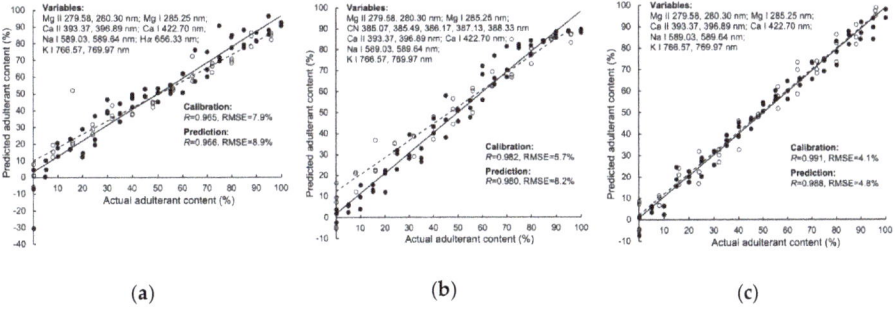

**Figure 3.** Scatter plot of actual adulterant content vs. LIBS measured adulterant content. Quantification of adulterant content in the mixture of (**a**) acacia honey and HFCS F55 based on the SR-PLSR model; (**b**) acacia honey and HFCS 90 based on the VIP-PLSR model; (**c**) acacia honey and rape honey based on the VIP-PLSR model.

## 4. Conclusions

In this study, LIBS combined with chemometric methods was used to detect honey adulteration. The adulterant content of acacia honey (adulterated with HFCS 55, HFCS 90, and rape honey) was successfully quantified. SR and VIP methods detected effectively the most relevant variables for adulteration determination. The emissions from Mg II 279.58, 280.30 nm, Mg I 285.25 nm, Ca II 393.37, 396.89 nm, Ca I 422.70 nm, Na I 589.03, 589.64 nm, and K I 766.57, 769.97 nm were considered as feature variables and played an important role in modeling. The importance of these variables was also verified in univariate analysis. The SR-PLSR, VIP-PLSR, and VIP-PLSR achieved the best results for detecting an adulteration ratio of HFCS F55, HFCS 90, and rape honey, with RMSE of 8.9%, 8.2%, and 4.8%, respectively. The results indicated the promising possibility of using LIBS and chemometric methods for quantification in honey adulteration. In addition, some research concerning model transfer could be explored, and more types of acacia honey as well as adulterants could be included in modeling in further study, which might be helpful for practical application.

**Author Contributions:** Conceptualization F.Z. and F.L.; Data curation, J.P.; Formal analysis, J.P.; Funding acquisition, J.P. and Z.Z.; Investigation, W.X.; Methodology, J.P.; Project administration, F.Z. and F.L.; Resources, J.J. and Z.Z.; Software, W.X.; Validation, W.X., F.Z. and F.L.; Writing – original draft, J.P.; Writing – review & editing, J.P., W.X., J.J., Z.Z., F.Z. and F.L. All authors have read and agreed to the published version of the manuscript.

**Funding:** This research was funded by National Key R&D Program of China, grant number 2018YFD0700502, Zhejiang Provincial Key Research and Development Program, grant number 2017C02027, And China Postdoctoral Science Foundation, grant number 2019M652143.

**Conflicts of Interest:** The authors declare no conflict of interest.

## References

1. Wu, L.; Du, B.; Vander Heyden, Y.; Chen, L.; Zhao, L.; Wang, M.; Xue, X. Recent advancements in detecting sugar-based adulterants in honey–A challenge. *Trends Anal. Chem.* **2017**, *86*, 25–38. [CrossRef]
2. Zhang, W.; Xue, J. Economically motivated food fraud and adulteration in China: An analysis based on 1553 media reports. *Food Control* **2016**, *67*, 192–198. [CrossRef]
3. Kendall, H.; Clark, B.; Rhymer, C.; Kuznesof, S.; Hajslova, J.; Tomaniovab, M.; Breretonc, P.; Frewera, L. A systematic review of consumer perceptions of food fraud and authenticity: A European perspective. *Trends Food Sci. Technol.* **2019**, *94*, 79–90. [CrossRef]
4. Amiry, S.; Esmaiili, M.; Alizadeh, M. Classification of adulterated honeys by multivariate analysis. *Food Chem.* **2017**, *224*, 390–397. [CrossRef] [PubMed]
5. Arroyo-Manzanares, N.; Garcia-Nicolas, M.; Castell, A.; Campillo, N.; Vinas, P.; Lopez-Garcia, I.; Hernandez-Cordoba, M. Untargeted headspace gas chromatography-Ion mobility spectrometry analysis for detection of adulterated honey. *Talanta* **2019**, *205*. [CrossRef]
6. Gan, Z.; Yang, Y.; Li, J.; Wen, X.; Zhu, M.; Jiang, Y.; Ni, Y. Using sensor and spectral analysis to classify botanical origin and determine adulteration of raw honey. *J. Food Eng.* **2016**, *178*, 151–158. [CrossRef]
7. Dramicanin, T.; Ackovic, L.L.; Zekovic, I.; Dramicaninn, M.D. Detection of Adulterated Honey by Fluorescence Excitation-Emission Matrices. *J. Spectrosc.* **2018**. [CrossRef]
8. Ferreiro-Gonzalez, M.; Espada-Bellido, E.; Guillen-Cueto, L.; Palma, M.; Barroso, C.G.; Barbero, G.F. Rapid quantification of honey adulteration by visible-near infrared spectroscopy combined with chemometrics. *Talanta* **2018**, *188*, 288–292. [CrossRef] [PubMed]
9. Jose Aliano-Gonzalez, M.; Ferreiro-Gonzalez, M.; Espada-Bellido, E.; Palma, M.; Barbero, G.F. A screening method based on Visible-NIR spectroscopy for the identification and quantification of different adulterants in high-quality honey. *Talanta* **2019**, *203*, 235–241. [CrossRef] [PubMed]
10. Zorov, N.B.; Popov, A.M.; Zaytsev, S.M.; Labutin, T.A. Qualitative and quantitative analysis of environmental samples by laser-induced breakdown spectrometry. *Russ. Chem. Rev.* **2015**, *84*, 1021–1050. [CrossRef]
11. Akpovo, C.A.; Martinez, J.A.; Lewis, D.E.; Branch, J.; Schroeder, A.; Edington, M.D.; Johnson, L. Regional discrimination of oysters using laser-induced breakdown spectroscopy. *Anal. Methods* **2013**, *5*, 3956–3964. [CrossRef]

12. Yang, P.; Zhou, R.; Zhang, W.; Yi, R.; Tang, S.; Guo, L.; Hao, Z.; Li, X.; Lu, Y.; Zeng, X. High-sensitivity determination of cadmium and lead in rice using laser-induced breakdown spectroscopy. *Food Chem.* **2019**, *272*, 323–328. [CrossRef] [PubMed]
13. Bilge, G.; Boyaci, I.H.; Eseller, K.E.; Tamer, U.; Cakir, S. Analysis of bakery products by laser-induced breakdown spectroscopy. *Food Chem.* **2015**, *181*, 186–190. [CrossRef] [PubMed]
14. Ferreira, E.C.; Menezes, E.A.; Matos, W.O.; Milori, D.M.B.P.; Nogueira, A.R.A.; Martin-Neto, L. Determination of Ca in breakfast cereals by laser induced breakdown spectroscopy. *Food Control* **2010**, *21*, 1327–1330. [CrossRef]
15. Lastra-Mejías, M.; Izquierdo, M.; González-Flores, E.; Cancilla, J.C.; Izquierdo, J.G.; Torrecilla, J.S. Honey exposed to laser-induced breakdown spectroscopy for chaos-based botanical classification and fraud assessment. *Chemom. Intellig. Lab. Syst.* **2020**, *199*. [CrossRef]
16. Liu, F.; Ye, L.; Peng, J.; Song, K.; Shen, T.; Zhang, C.; He, Y. Fast detection of copper content in rice by laser-induced breakdown spectroscopy with uni- and multivariate analysis. *Sensors* **2018**, *18*, 705. [CrossRef]
17. Haaland, D.M.; Thomas, E.V. Partial Least-Squares Methods for Spectral Analyses. 1. Relation to Other Quantitative Calibration Methods and the Extraction of Qualitative Information. *Anal. Chem.* **1988**, *60*, 1193–1202. [CrossRef]
18. Jong, S.D. SIMPLS: An alternative approach to partial least squares regression. *Chemom. Intellig. Lab. Syst.* **1993**, *18*, 251–263. [CrossRef]
19. Forrest, S. Genetic algorithms: Principles of natural selection applied to computation. *Science* **1993**, *261*, 872–878. [CrossRef]
20. Farrés, M.; Platikanov, S.; Tsakovski, S.; Tauler, R. Comparison of the variable importance in projection (VIP) and of the selectivity ratio (SR) methods for variable selection and interpretation. *J. Chemom.* **2015**, *29*, 528–536. [CrossRef]
21. Kramida, A.; Ralchenko, Y.; Reader, J. *Atomic Spectra Database*; National Institute of Standards and Technology: Gaithersburg, MD, USA, 2015.
22. Fu, X.; Duan, F.; Huang, T.; Ma, L.; Jiang, J.; Li, Y. A fast variable selection method for quantitative analysis of soils using laser-induced breakdown spectroscopy. *J. Anal. At. Spectrom.* **2017**, *32*, 1166–1176. [CrossRef]
23. Lu, S.; Shen, S.; Huang, J.; Dong, M.; Lu, J.; Li, W. Feature selection of laser-induced breakdown spectroscopy data for steel aging estimation. *Spectrochim. Acta Part B* **2018**, *150*, 49–58. [CrossRef]

© 2020 by the authors. Licensee MDPI, Basel, Switzerland. This article is an open access article distributed under the terms and conditions of the Creative Commons Attribution (CC BY) license (http://creativecommons.org/licenses/by/4.0/).

Article

# Oilseeds from a Brazilian Semi-Arid Region: Edible Potential Regarding the Mineral Composition

Ivone M. C. Almeida [1], M. Teresa Oliva-Teles [2], Rita C. Alves [1], Joana Santos [1], Roberta S. Pinho [3], Suzene I. Silva [3], Cristina Delerue-Matos [2] and M. Beatriz P. P. Oliveira [1,*]

[1] REQUIMTE/LAQV, Departamento de Ciências Químicas, Faculdade de Farmácia, Universidade do Porto, Rua Jorge Viterbo Ferreira, 228, 4050–313 Porto, Portugal; ivonemalmeida@gmail.com (I.M.C.A.); rcalves@ff.up.pt (R.C.A.); joanasantoscma@sapo.pt (J.S.)
[2] REQUIMTE/LAQV, Instituto Superior de Engenharia do Porto, Instituto Politécnico do Porto, Rua Dr. António Bernardino de Almeida, 431, 4200–072 Porto, Portugal; mtt@isep.ipp.pt (M.T.O.-T.); cmm@isep.ipp.pt (C.D.-M.)
[3] Laboratório de Recursos Econômicos e Fitoquímica, Departamento de Biologia, Área de Botânica, Universidade Federal Rural de Pernambuco, Av. Dom Manoel de Medeiros s/n, Dois Irmãos, CEP, Recife-PE 52171–900, Brazil; rsampaiop@gmail.com (R.S.P.); suzene@db.ufrpe.br (S.I.S.)
* Correspondence: beatoliv@ff.up.pt

Received: 17 January 2020; Accepted: 18 February 2020; Published: 21 February 2020

**Abstract:** Oilseeds from five native plant species with edible potential from the Brazilian Caatinga semi-arid region (*Diplopterys pubipetala*, *Barnebya harleyi*, *Croton adamantinus*, *Hippocratea volubilis*, and *Couroupita guianensis*) were investigated regarding their mineral contents. The minerals, Na, K, Ca, Mg, Fe, Cu, Cr, Al, were analyzed by high-resolution continuum source atomic absorption spectrometry (HR–CS AAS) and P by the vanadomolybdophosphoric acid colorimetric method. K, Mg, and P were the main elements found (1.62–3.7 mg/g, 362–586 µg/g, and 224–499 µg/g dry weight (dw), respectively). *B. harley* seeds contained the highest amounts of K and P, while *C. guianensis* seeds were the richest in Mg. Fe was the most abundant oligoelement (2.3–25.6 µg/g dw). Cr contents were below the limit of quantification for all samples and Al amounts were low: 0.04–1.80 µg/g dw. A linear discriminant analysis clearly differentiated *B. harleyi* and *C. guianensis* samples from the remaining ones. In sum, these oilseeds from the Brazilian Caatinga semi-arid region seem to have the potential to be used as natural sources of minerals, mainly K.

**Keywords:** oilseeds; Caatinga; native; minerals; spectrometry

## 1. Introduction

The plant species that grow in the arid and semi-arid land regions of the world have attracted considerable attention in recent years. The Caatinga, the main ecosystem in the Northeastern Brazil (~800.000 km$^2$), is a seasonally dry tropical forest composed by a heterogeneous mix of plant species, primarily deciduous xerophytic spiny shrubs, and trees [1,2]. Despite the huge diversity, only a few of the species of this semi-arid region are exploited for industrial purposes [2]. The rural populations living in this area use many of these plants for food, fuel, timber, medicines, and livestock feed. These products contribute to income generation and offer an alternative source of livelihood when crops fail under the erratic and low rainfall conditions. However, human intervention over the years, such as deforestation, livestock pasture, and cropping, has contributed to the loss of vegetation cover and biodiversity, and extensive soil degradation in this biome [3–9].

Although different studies have already identified some plant species with great economic potential [2,3,8–11], the lack of scientific knowledge for the majority remains. The evaluation of the chemical composition of non-conventional tropical plants is essential not only for their fundamental knowledge, but also to support their use as raw material for food, chemical, and pharmaceutical industries [5]. For instance, in a recent study, Andrade et al. showed that native species from Caatinga used by the local population to treat inflammatory disorders have also a good photoprotective potential and could be used for pharmaceutical preparations [11]. In addition, nuts and seeds from the tropics have been shown to be rich in lipids and relatively good sources of inexpensive and renewable carbohydrates and proteins, and compounds with high added value, like lipid-soluble vitamins, phytosterols, and other phytochemicals, as well as notable quantities of minerals [12–14].

The insufficiency of minerals in humans is mainly caused by an unbalanced diet. In fact, Fe, Ca, Mg, and Cu are among the mineral elements most commonly lacking in human diets [15], and their deficiency could have a negative impact on health. The information about the mineral composition of unconventional oilseeds from Caatinga is very scarce. The aim of this work was, therefore, to analyze the mineral profile of oilseeds from five native species of this region using high-resolution continuum source atomic absorption spectrometry (HR–CS AAS), in order to evaluate their edible potential as a source of minerals. The seeds selected for this study are from Brazilian plants traditionally used in folk medicine, namely, *Croton adamantinus* Mull. Arg. (Euphorbiaceae), *Barnebya harleyi* W.R. Anderson & B. Gates *Diplopterys pubipetala* (A. Juss.) W.R. Anderson & C. Davis, (Malpighiaceae), *Hippocratea volubilis* L. (Celastraceae), and *Couroupita guianensis* Aubl. (Lecythidaceae). The present work is, as far as we know, the first published report on the mineral composition of these oilseeds.

## 2. Materials and Methods

### 2.1. Chemicals and Standard Solutions

Ultrapure water from a Simplicity 185 system (resistivity 18.2 MΩ cm; Millipore, Belford, USA) and nitric acid (65%; Suprapure®, Merck, Darmstadt, Germany) were used for the preparation of samples and standards. All reagents were of analytical or suprapure reagent-grade. Working standard solutions were prepared from the stock solutions of Na, K, Ca, Mg, Fe, Cu, Cr, and Al (1000 mg/L; Panreac, Barcelona, Spain) by proper dilution with a 0.5% and 1% (v/v) nitric acid solution. Cesium chloride (0.1% w/v; Sigma-Aldrich, Steinheim, Germany) was used as flame ionization chemical suppressor. For Ca analyses, lanthanum chloride heptahydrate (1%; w/v) was added to all solutions in order to minimize the oxy–salts interferences in FAAS analysis. A 0.1% (w/v) amount of magnesium nitrate (10 g/L; Panreac, Barcelona, Spain) was used as matrix modifier for the electrothermal atomic absorption spectrometry (ETAAS) quantification of Fe and Al; 0.05% (w/v) magnesium and 0.1% (w/v) palladium nitrates (10 g/L; Merck, Darmstadt, Germany) for Cu analysis.

Potassium dihydrogen phosphate (99.5%; Riedel-de Haën, Seelze, Germany), ammonium heptamolybdate tetrahydrated (99.0%, Merck, Darmstadt, Germany), and ammonium monovanadate (99.0%, Merck, Darmstadt, Germany) were the reagents used for P analysis. Working standards solutions were manually prepared for HR–CS FAAS and P analysis. Solutions for HR–CS ETAAS calibration curves were automatically prepared by the autosampler.

All glassware and plastic materials were cleaned by treatment with nitric acid (10%, v/v) for 24 h and then rinsed with water, prior to use.

### 2.2. Samples

Mature fruits of five different indigenous plant species (*Diplopterys pubipetala*, *Barnebya harleyi*, *Croton adamantinus*, *Hippocratea volubilis*, and *Couroupita guianensis*) were obtained from five individuals of each species in Pernambuco State (Brazil).

The fruits were dehydrated at 60 °C for 48 h, until constant weight, and seeds were manually removed from the fruits. Seeds of each species were thinly ground in a mill and stored in plastic containers at 4 °C until use.

## 2.3. Sample Preparation

Aliquots (1 g) of the dried ground seed specimens were dry-ashed in a muffle furnace at 500 °C, overnight. To the resulting white ash, 200 µL of nitric acid (65%) was added and ashed for another hour at 500 °C. An amount of 5.0 mL of nitric acid (65%) was used to re-dissolve minerals and, afterward, evaporated to a final volume of 2 mL. After filtration, the solutions were diluted to 50 mL with ultrapure water. Blanks were also carried out. The mineralization procedure was done in triplicate for each composite sample [16].

## 2.4. Analytical Methodologies

An Analytik Jena contrAA 700 atomic absorption spectrometer (Analytik Jena, Germany), equipped with a high-intensity xenon short-arc lamp as continuum radiation source, and automatic MPE60 and AS52s autosamplers (Analytik, Jena, Germany), for electrothermal and flame atomization, respectively, was used throughout this work. Air-acetylene flame (high-purity grade) was used for K, Mg, Ca, and Na quantification at 766.4908 nm, 285.2125 nm, 422.6728 nm, and 588.9953 nm, respectively. Transversely heated graphite furnace with platform-pyrolytically coated graphite tubes (Analytik Jena, Germany) and high-purity argon was used for electrothermal determinations of Fe (248.3270 nm), Cu (324.7540 nm), Cr (357.8687 nm), and Al (309.2713 nm). Pyrolysis and atomization temperatures were optimized to maximize absorbance signals and to minimize backgrounds and matrix interferences. The optimized electrothermal programs used are shown in Table 1. Measurements were performed at the selected primary atomic lines at a wavelength-integrated absorbance equivalent to 3 pixels (central pixel ± 1).

**Table 1.** Temperature programs for Fe, Cu, Cr, and Al determination in seeds by high-resolution continuum source electrothermal atomic absorption spectrometry (HRCS ETAAS).

| | Temperature (Ramp Time, Hold Time); °C (°C/s, s) | | | |
|---|---|---|---|---|
| - | Fe | Cu | Cr | Al |
| Drying | 90 (3, 23.3) | 90 (3, 23.3) | 90 (3, 23.3) | 90 (3, 23.3) |
| Pyrolysis | 950 (300, 12.0) | 900 (300, 11.8) | 1300 (300, 13.2) | 1400 (300, 13.5) |
| Atomization | 1900 (1500, 4.6) | 1800 (1500, 4.6) | 2400 (1500, 4.7) | 2450 (1500, 3.7) |
| Cleaning | 2450 | 2450 | 2450 | 2500 |

The method accuracy was assessed through recovery experiments of the seed samples spiked with three different concentrations (50%, 100%, and 125% of the sample contents) of the aqueous standards. Precision was evaluated through intra- and inter-day measurements, by performing three sample measurements on the same day (intra-day) and over three days (inter-day). The limits of quantification (LOQ) were based on the residual standard deviation (10-fold) of calibration curves.

P was determined by the vanadomolybdophosphoric acid colorimetric method according to the Standard Method 4500-P [16]. The color development reagent for P determination was prepared by the addition of ammonium heptamolybdate tetrahydrated and ammonium monovanadate. Phosphorus measurements were performed at 420 nm in a dual-beam UV/Vis spectrophotometer (Evolution 300, Thermo Scientific, Boston, MA, USA).

## 2.5. Statistical Analysis

The results were reported as mean ± standard deviation. One-way analysis of variance (ANOVA) and Tukey's HSD post-hoc test were applied to evaluate the possible significant differences ($p < 0.05$) in mineral content between the different seeds. The mineral compositions of the seeds were also compared using multivariate statistical analysis. A linear discriminant analysis (LDA) was performed with forward stepwise analysis, considering a $p$-value of 0.01, using Wilk's lambda ($\lambda$) as the selection criterion, and an F-statistic value to determine the significance of the changes $\lambda$ when a new variable is tested. Followed this was canonical analysis to retrieve more information about the nature of the discrimination between the samples. Statistical analysis was carried out using STATISTICA software v.10 for Windows (Statsoft Inc., Tulsa, USA).

## 3. Results and Discussion

In previous research, Pinho et al. already identified *D. pubipetala*, *B. harleyi*, *C. adamantinus*, and *H. volubilis* as promising species for cultivation, based on their oil content and fatty acids profile [13]. *C. adamantinus* is used in the treatment of wounds, due to its anti-inflammatory and analgesic properties. Their seeds contain 37.1% of oil, with high concentrations of essential linoleic (44.2%) and linolenic (45.2%) fatty acids, and are rich in vitamin E, interesting characteristics for the human diet [13,17]. *B. harleyi* and *D. pubipetala* seeds contain high amounts of protein and are promising sources of oil (46.4% and 45.3%, respectively), in particular, of α-linolenic acid (42.8% and 31.9%, respectively) [13]. *H. volubilis* is a traditional expectorant. The seeds contain edible oil (45%–50%), but with a slightly bitter taste. *C. guianensis* leaves, flowers, and barks are known to contain many active ingredients that possess antibiotic, antifungal, antiseptic, antinociceptive, and immunostimulant activities ([18–20]. They contain about ~30% of oil, consisting mainly in linoleic acid (>80%) [13,21,22].

The minerals analyzed in this work were selected based on their physiological importance (e.g., Ca, Mg, K, ... ) or relevance as eventual indicators of environmental contamination (e.g., Cr and Al), since the perception of their composition is essential to better understand their potential to be used in human and/or animal nutrition or, for instance, in the dietary supplement industry.

Flame (FAAS) and electrothermal atomic absorption spectrometry (ETAAS), inductively coupled plasma optical emission spectrometry (ICP–OES), and inductively coupled plasma mass spectrometry (ICP–MS) are the main methods employed in mineral analysis [23]. In this work, the more recent technique HR–CS AAS was employed to evaluate the mineral content of the referred seeds. This equipment presents main advantages over the traditional line source AAS, namely the visualization of the spectral environment in the vicinity of the analytical line, the simultaneous background correction, and the improvement in precision and detection limits due to the higher signal-to-noise ratio [24,25].

### 3.1. Method Validation

Method parameters including linear range and correlation coefficient ($r$), LOQ values, repeatability, and mean recovery percentages are presented in Table 2. Na, K, Ca, Mg, P, Fe, Cu, Cr, and Al were quantified in seeds by the external calibration method with aqueous standard solutions. Good linearity was achieved for all elements over the range of tested concentrations, with correlation coefficients higher than 0.997 for all calibration plots. The LOQ ranged from 4.6 (Ca) and 11 µg/g (Ca) for major elements, and from 0.025 (Cu) to 0.11 µg/g (Al) for minor elements. Accuracy and precision of the method were good: recoveries ranged from 85% to 114%, intra-day precisions from 1.5% to 5.2%, and inter-day relative standard deviation (RSD) between 2.1% and 7.9%. These data confirmed the suitability of the method for the designated application.

**Table 2.** Methods validation parameters: linear range and correlation coefficients (r) of the analytical calibration curves, intra- and inter-day precisions (RSD, %), and limits of quantification (LOQ) for the elements studied.

| Element | Wavelength (nm) | Detection | Calibration | r | LOQ (µg/g) | Precision RSD [a] (%) Intra-Day | Inter-Day | Recovery [b] (%) |
|---|---|---|---|---|---|---|---|---|
| Na | 588.9953 | HR-CS FAAS | 0.1–1 (mg/L) | 0.9992 | 8.4 | 1.6 | 2.4 | 85 |
| K | 766.4908 | HR-CS FAAS | 0.1–1 (mg/L) | 0.9992 | 8.6 | 1.5 | 2.3 | 112 |
| Ca | 422.6728 | HR-CS FAAS | 0.1–1 (mg/L) | 0.9996 | 4.6 | 3.5 | 4.3 | 112 |
| Mg | 285.2125 | HR-CS FAAS | 0.1–1 (mg/L) | 0.9997 | 6.5 | 3.4 | 4.2 | 103 |
| P | 420.0 | UV/Vis | 0.1–3 (mg/L) | 0.9995 | 11 | 1.6 | 2.1 | 104 |
| Fe | 248.3270 | HR-CS ETAAS | 3–15 (µg/L) | 0.9976 | 0.090 | 5.1 | 6.3 | 93 |
| Cu | 324.7540 | HR-CS ETAAS | 3–15 (µg/L) | 0.9999 | 0.025 | 3.7 | 4.3 | 106 |
| Cr | 357.8687 | HR-CS ETAAS | 3–15 (µg/L) | 0.9998 | 0.039 | 5.2 | 7.9 | 110 |
| Al | 309.2713 | HR-CS ETAAS | 0.1–40 (µg/L) | 0.9999 | 0.11 | 3.4 | 4.8 | 114 |

[a] Average RSD: inter-day (n = 3); intra-day (n = 3, three days).; [b] Mean recovery of experiments with three levels of sample spiking (50%, 100%, and 125% of the sample contents).

## 3.2. Seed Analysis

The potential of the selected oilseeds as sources of minerals (Na, K, Ca, Mg, P, Fe, Cu, Cr) was ascertained and is described in Table 3.

**Table 3.** Mineral concentrations (µg/g dry weight (dw)) of the analyzed oilseeds.

| Element | Species | | | | |
|---|---|---|---|---|---|
| | Diplopterys pubipetala | Barnebya harleyi | Croton adamantinus | Hippocratea volubilis | Couroupita guianensis |
| Na | 38 ± 1 [c] | 106 ± 6 [a] | 90 ± 12 [ab] | 74 ± 14 [b] | 8 ± 1 [d] |
| K | 1636 ± 135 [c] | 3649 ± 215 [a] | 1719 ± 282 [c] | 2554 ± 89 [b] | 1732 ± 187 [c] |
| Ca | 134 ± 6 [c] | 134 ± 9 [c] | 147 ± 14 [cb] | 170 ± 2 [b] | 227 ± 12 [a] |
| Mg | 447 ± 13 [b] | 494 ± 39 [b] | 362 ± 0.4 [c] | 384 ± 30 [c] | 586 ± 12 [a] |
| P | 288 ± 55 [c] | 499 ± 30 [a] | 397 ± 1 [b] | 224 ± 36 [c] | 419 ± 4 [ab] |
| Fe | 3.69 ± 0.03 [c] | 25.6 ± 2.5 [a] | 2.34 ± 0.23 [c] | 6.88 ± 0.19 [b] | 7.50 ± 0.54 [b] |
| Cu | <LOQ | 0.77 ± 0.15 [a] | 0.046 ± 0.002 [b] | <LOQ | 0.043 ± 0.005 [b] |
| Cr | <LOQ | <LOQ | <LOQ | <LOQ | <LOQ |
| Al | <LOQ | 1.00 ± 0.14 [b] | <LOQ | 0.192 ± 0.017 [c] | 1.80 ± 0.08 [a] |

Values are presented as mean ± standard deviation of triplicate determinations. Within each line, different letters represent significant differences between species, at $p < 0.05$.

Physiologically, the macroelements Na, K, Ca, and Mg are essential for several human body functions, while P is the principal reservoir for metabolic energy and a co-factor of many enzymes [26]. K was the most prevalent element in the mineral composition of these seeds, followed, in decreasing order, by Mg, P, Ca, and Na. Among the studied samples, B. harleyi contained the highest K concentration (3649 µg/g). K contents ranged from 1636 to 3649 µg/g, levels considerably higher than those reported by Onyeike and Acheru [27] for castor seeds, coconut seeds, dikanut seeds, groundnut seeds, melon seeds, oil bean seeds, and palm kernel seeds (134–168 µg/g). This is an interesting characteristic taking into account the health importance of K.

The Mg contents in the samples (362 to 586 µg/g) were very modest compared to the levels found for some common culinary nuts and seeds (almonds (~2000 µg/g), pecans (~1500 µg/g), hazelnuts (2000 µg/g) [28], and mustard oilseeds (~7000 µg/g) [29].

The P content ranged from 224 to 499 µg/g. These results are lower than those reported for umbu (*Spondias tuberosa* Arr. Cam) seeds (7720–8250 µg/g) [10] and baru (*Dipteryx alata* Vog.) almonds (7300 µg/g) [30], two Brazilian plant species.

Ca levels (134 to 227 µg/g) were lower than those reported for pumpkin seeds (390–420 µg/g), sunflower seeds, and pistachios (~900 µg/g), and peanuts (520–530 µg/g) [28], but seeds from *D. pubipetala*, *B. harleyi*, and *H. volubilis* presented Ca values identical to pine nuts (130 µg/g) [28]. The richest source of Ca were the seeds of *C. guianensis* (227 µg/g), with levels slightly higher to those found in the nuts of *Cyperus esculentus* (188 µg/g) [31].

The Na content was low for all the samples (<106 µg/g), another good characteristic for edible seeds having in view its health effects.

Fe and Cu are essential trace elements. About two-thirds of the total body iron are involved in metabolic or enzymatic functions, mostly in erythrocytes (hemoglobin) and muscles (myoglobin) while the physiological functions of Cu arise directly from its role in a number of Cu containing enzymes, being essential for brain growth and maturation. Cr is also an essential element usually present in food in the trivalent form. In living organisms, Cr (III) is a component of enzymes that controls glucose metabolism and synthesis of fatty acids and cholesterol. Cr (VI) is toxic and carcinogenic to humans, but it is not usually present in food [32,33].

From a food safety perspective, Al contents of the oilseeds were also evaluated. Most of the population is not at risk for Al toxicity since humans have diverse mechanisms to prevent significant absorption and to aid its elimination. However, when protective gastrointestinal mechanisms are bypassed, renal function is impaired or exposure is high, Al can accumulate in the body, and originate impaired neurological development, Alzheimer's disease, metabolic bone disease, dyslipidemia or even genotoxic activity [34].

The levels obtained for the minor minerals Cu, Cr, and Al were generally low for all samples. Fe was the most abundant of the analyzed oligoelements in all seeds, ranging from 2.34 µg/g in *C. adamantinus* to ten-fold more (25.6 µg/g) in *B. harleyi*. These values are lower than those described for mustard seeds (75–170 µg/g) [29], but the Fe level in *B. harleyi* is comparable to the values found in the literature for some well-known edible nuts such as pecan nuts (24–26 µg/g) and peanuts (20 µg/g) [28], and to the tropical babassu (*Orbignya speciosa*) nuts (18–33 µg/g) and sapucaia (*Lecythis pisonis*) nuts (21–36 µg/g) [14]. *H. volubilis* and *C. guianensis* oilseeds contained similar Fe concentrations to those reported for cupuassu (*Theobroma grandiflorum*) seeds (7.5 µg/g) [14]. *D. pubipetala* and *H. volubilis* seeds contained less than quantifiable levels of Cu (<0.025 µg/g) and the remaining samples contained lower amounts than those found in mustard seeds (6–13 µg/g). Cr concentrations fell below its quantification limit (<0.039 µg/g) in all seeds. Aluminum concentrations ranged from 0.044 µg/g in *D. pubipetala* to 1.80 µg/g in *C. guianensis* seeds, less than the concentration indicated for *Allantoma lineata* seeds (8 µg/g), an Amazon tree from the same family [22].

In sum, *B. harley* seeds had the highest content of K, P, Na, Fe, and Cu, while *C. guianensis* presented the highest levels of Mg, Ca, and Al from the five analyzed seeds.

Although all the oilseeds analyzed showed, in general, a similar mineral profile (high concentration of K followed by Mg > P > Ca > Na > Fe > Al > Cu), their composition showed several significant ($p < 0.05$) differences between them (Table 3). A Linear Discriminant Analysis (LDA) was performed to determine which minerals were able to discriminate between two or more naturally occurring groups (Table 4).

Table 4. Summary of discriminant function analysis.

| - | Wilks' Lambda | Partial Lambda | F-Remove (4,4) | p-Level | Toler. |
|---|---|---|---|---|---|
| Na | 0.000 | 0.163 | 3.840 | 0.149 | 0.173 |
| K | 0.000 | 0.302 | 1.737 | 0.339 | 0.333 |
| Ca | 0.000 | 0.205 | 2.910 | 0.203 | 0.290 |
| Mg | 0.000 | 0.557 | 0.597 | 0.692 | 0.289 |
| P | 0.000 | 0.391 | 1.168 | 0.468 | 0.732 |
| Fe | 0.000 | 0.094 | 7.270 | 0.067 | 0.069 |
| Cu | 0.000 | 0.125 | 5.265 | 0.102 | 0.106 |
| Al | 0.000 | 0.016 | 45.689 | 0.005 | 0.056 |

Wilks' Lambda: 0.00000 approx. F (28.15) = 56.326, $p < 0.0000$.

The model was generated through forward stepwise analysis, considering variables one by one, choosing those with the most significant contribution to the data discrimination. Cr content was not considered in this multivariate analysis, as it was always inferior to the LOQ of the analytical method. All the other minerals (K, Mg, P, Ca, Na, Fe, Cu, and Al) were selected for the model generated by the LDA, Al and Fe being the variables that showed the highest contribution to the model. A subsequent canonical analysis generated, in total, four canonical variates (CV). The minerals and the case samples were displayed in a canonical variate scatterplot (Figure 1) of the first two discriminant functions, which comprised 98.9% of the information contained in the generated model. The first dimension (CV 1) represents 70.0% of the data variance, being more related to their Al, Fe, and Cu content. The second canonical variate (CV 2) adds more than 28.9% of information to the model, separating the samples according to their Fe, K, Ca, and Mg content. The relations of the variables with the first two canonical variates are presented by the resulting equations:

$$CV\ 1 = -4.18[Al] + 3.34[Fe] - 0.97[Mg] - 0.048[P] + 1.29[Ca] - 2.65[Cu] - 1.73[K] - 0.81[Na] \quad (1)$$

$$CV\ 2 = -0.15[Al] + 1.19[Fe] - 0.73[Mg] + 0.13[P] + 0.83[Ca] + 1.06[Cu] + 1.23[K] + 0.43[Na] \quad (2)$$

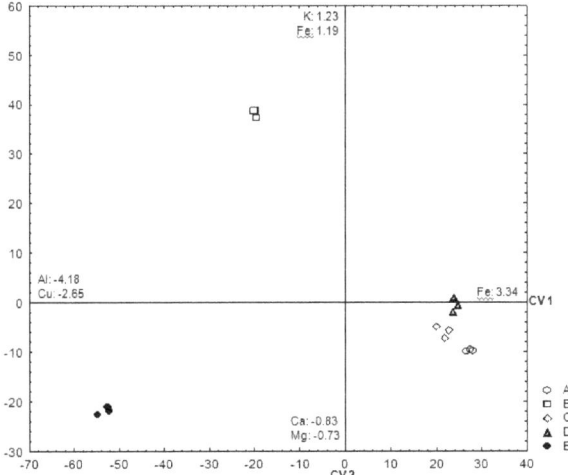

Figure 1. Scatterplot of all analyzed cases on the first canonical variate (CV1) versus the second canonical variate (CV2) (A: *Diplopterys pubipetala*; B: *Barnebya harleyi*; C: *Croton adamantinus*; D: *Hippocratea volubilis*; E: *Couroupita guianensis*).

From the graphical display of the studied cases (see Figure 1), there were two seed species, namely *B. harleyi* and *C. guianensis*, visibly separated from the others. In the CV1, the Al content of the samples allowed to clearly distinguish the separation between the *C. guianensis* seeds and the others. The *B. harleyi* seeds were discriminated by their highest content of Fe, K, and Al content. The seeds *D. pubipetala*, *C. adamantinus*, and *H. volubilis* were displayed more closely to the positive side of the CV1 showing that the mineral composition of these three seeds was more similar.

## 4. Conclusions

The mineral contents of oilseeds from five native plant species from Brazil were evaluated. All the samples analyzed seem to be good sources of K, Mg, and P and presenting also low levels of Na. Nevertheless, *B. harley* oilseeds presented significantly higher amounts ($p < 0.05$) of K and P, Fe, and Cu compared with the other species. *C. guianensis* seeds were the best source of Mg. These two samples were, indeed, clearly differentiated from the remaining ones through linear discriminant analysis. In general, Cr and Al contents were low for all samples. The results show that these oilseeds from the Brazilian Caatinga semi-arid region have the potential to be used as food products and as mineral sources (mainly K).

**Author Contributions:** Conceptualization, R.S.P., S.I.S. and M.B.P.P.O.; Formal analysis, I.M.C.A., R.C.A. and J.S.; Funding acquisition, C.D.-M., M.B.P.P.O.; Investigation, I.M.C.A. and M.T.O.-T.; Methodology, M.T.O.-T.; Project administration, S.I.S. and M.B.P.P.O.Resources, R.S.P., S.I.S. and C.D.-M.; Supervision, M.T.O.-T., C.D.-M. and M.B.P.P.O.; Validation, M.T.O.-T.; Writing-original draft, I.M.C.A., R.C.A. and J.S.; Writing-review & editing, M.T.O.-T. and R.C.A.

**Funding:** The work was supported by UIDB/50006/2020 with funding from FCT/MCTES through national funds.

**Acknowledgments:** R. C. Alves is grateful to *Fundação para a Ciência e a Tecnologia* for the CEECIND/01120/2017 contract.

**Conflicts of Interest:** The authors declare no conflicts of interest.

## References

1. Araújo, E.L.; Castro, C.C.; Albuquerque, U.P. Dynamics of Brazilian Caatinga-A review concerning the plants, environment and people. *Funct. Ecosystems. Commun.* **2007**, *1*, 15–28.
2. Silva, R.B.; Silva-Júnior, E.V.; Rodrigues, L.C.; Andrade, L.H.C.; Silva, S.I.; Harand, W.; Oliveira, A.F. A comparative study of nutritional composition and potential use of some underutilised tropical fruits of Arecaceae. *An. Acad. Bras. Ciênc.* **2015**, *87*, 1701–1709. [CrossRef] [PubMed]
3. Crepaldi, I.C.; Almeida-Muradian, L.B.D.; Rios, M.D.G.; Penteado, M.D.V.C.; Salatino, A. Composição nutricional do fruto de licuri (*Syagrus coronata* (Martius) Beccari). *Br. J. Bot.* **2001**, *24*, 155–159. [CrossRef]
4. Lucena, R.F.P.; Albuquerque, U.P.; Monteiro, J.M.; de Almeida, C.F.; Florentino, A.T.; Ferraz, J.S. Useful plants of the semi-Arid Northeastern region of Brazil–A look at their conservation and sustainable use. *Environ. Monit. Assess.* **2007**, *125*, 281–290. [CrossRef] [PubMed]
5. Harand, W.; Pinho, R.S.; Felix, L.P. Alternative oilseeds for Northeastern Brazil: Unrevealed potential of Brazilian biodiversity. *Br. J. Bot.* **2016**, *39*, 169–183. [CrossRef]
6. Maia, L.; Leão, M.D.; Barbosa, M.; de Souza, S.; Coutinho, C.; Pastori, P. Entomofauna diversity in areas of Caatinga under forest management in the semi-arid region of Ceará. *Comun. Sci.* **2019**, *10*, 10–20. [CrossRef]
7. Albuquerque, U.P.; Andrade, L.H.C. Uso de recursos vegetais da Caatinga: O caso do agreste do Estado de Pernambuco (Nordeste do Brasil). *Interciencia* **2002**, *27*, 336–345.
8. Nascimento, V.T.; Moura, N.P.; Vasconcelos, M.A.S.; Maciel, M.I.S.; Albuquerque, U.P. Chemical characterization of native wild plants of dry seasonal forests of the semi-arid region of northeastern Brazil. *Food Res. Int.* **2011**, *44*, 2112–2119. [CrossRef]
9. Nascimento, V.T.; Vasconcelos, M.A.S.; Maciel, M.I.S.; Albuquerque, U.P. Famine foods of Brazil's seasonal dry forests: Ethnobotanical and nutritional aspects. *Econ. Bot.* **2012**, *66*, 22–34. [CrossRef]
10. Borges, S.V.; Maia, M.C.A.; Gomes, R.C.M.; Cavalcanti, N.B. Chemical composition of umbu (*Spondias tuberosa* Arr. Cam) seeds. *Quím Nova* **2007**, *30*, 49–52. [CrossRef]

11. Andrade, B.A.; Corrêa, A.J.C.; Gomes, A.K.S.; Neri, P.M.S.; Sobrinho, T.J.S.P.; Araújo, T.A.S.; Castro, V.T.N.A.; Amorim, E.L.C. Photoprotective activity of medicinal plants from the caatinga used as anti-Inflammatories. *Phcog. Mag.* **2019**, *15*, 356–361. [CrossRef]
12. Welna, M.; Klimpel, M.; Zyrnicki, W. Investigation of major and trace elements and their distributions between lipid and non-lipid fractions in Brazil nuts by inductively coupled plasma atomic optical spectrometry. *Food Chem.* **2008**, *111*, 1012–1015. [CrossRef]
13. Pinho, R.S.; Oliveira, A.F.M.; Silva, S.I. Potential oilseed crops from the semiarid region of northeastern Brazil. *Bioresour. Technol.* **2009**, *100*, 6114–6117. [CrossRef] [PubMed]
14. Naozuka, J.; Vieira, E.C.; Nascimento, A.N.; Oliveira, P.V. Elemental analysis of nuts and seeds by axially viewed ICP OES. *Food Chem.* **2011**, *124*, 1667–1672. [CrossRef]
15. White, P.J.; Brow, P.H. Plant nutrition for sustainable development and global health. *Ann. Bot.* **2010**, *105*, 1073–1080. [CrossRef]
16. APHA (American Public Health Organization). *Standard Methods for the Examination of Water and Wastewater*, 21th ed.; American Public Health Association: Washington DC, USA, 2005.
17. Silva, J.S.; Sales, M.F.; Carneiro-Torres, D.S. O gênero *Croton* (Euphorbiaceae) na microrregião do Vale do Ipanema, Pernambuco, Brasil. *Rodriguésia* **2009**, *60*, 879–901. [CrossRef]
18. Andrade, E.H.A.; Zoghbi, M.G.B.; Maia, J.G.S. The volatiles from flowers of *Couroupita guianensis* Aubl., *Lecythis usitata* Miers. var. paraensis (Ducke) R. Kunth. and *Eschweilera coriacea* (A. P. DC.) Mori (Lecythidaceae). *J. Essent. Oil Res.* **2000**, *12*, 163–166. [CrossRef]
19. Umachigi, P.S.; Jayaveera, K.N.; Kumar, C.K.A.; Kumar, S. Antimicrobial, wound healing and antioxidant potential of *Couroupita guianensis* in rats. *Pharmacologyonline* **2007**, *3*, 269–281.
20. Pinheiro, M.M.G.; Bessa, S.O.; Fingolo, C.E.; Kuster, R.M.; Matheus, M.E.; Menezes, F.S.; Fernandes, P.D. Antinociceptive activity of fractions from *Couroupita guianensis* Aubl. leaves. *J. Ethnopharmacol.* **2010**, *127*, 407–413. [CrossRef]
21. Lago, R.C.A.; Pereira, D.A.; Siqueira, F.A.R.; Szpiz, R.R.; Oliveira, J.P. Estudo preliminar das sementes e do óleo de cinco espécies da Amazônia. *Acta Amaz.* **1987**, *16*, 369–376. [CrossRef]
22. Andrade, E.H.A.; Maia, J.G.S.; Streich, R.; Marx, F. Seed composition of Amazonian Lecythidaceae species: Part 3 in the Series "Studies of edible Amazonian plants". *J. Food Comp. Anal.* **1999**, *12*, 37–51. [CrossRef]
23. Taylor, A.; Branch, S.; Halls, D.; Patriarca, M.; White, M. Atomic spectrometry update. Clinical and biological materials, foods and beverages. *J. Anal. Atomic. Spectrom.* **2005**, *20*, 323–369. [CrossRef]
24. Welz, B. High-Resolution continuum source AAS: The better way to perform atomic absorption spectrometry. *Anal. Bioanal. Chem.* **2005**, *381*, 69–71. [CrossRef] [PubMed]
25. Oliveira, S.R.; Raposo, J.L.; Gomes Neto, J.A. Fast sequential multi-element determination of Ca, Mg, K, Cu, Fe, Mn and Zn for foliar diagnosis using high-resolution continuum source flame atomic absorption spectrometry: Feasibility of secondary lines, side pixel registration and least-squares background correction. *Spectrochim. Acta Part B* **2009**, *64*, 593–596.
26. Stein, A.J. Global impacts of human mineral malnutrition. *Plant Soil* **2010**, *335*, 133–154. [CrossRef]
27. Onyeike, E.N.; Acheru, G.N. Chemical composition of selected Nigerian oil seeds and physicochemical properties of the oil extracts. *Food Chem.* **2002**, *77*, 431–437. [CrossRef]
28. Rodushkin, I.; Engström, E.; Sörlin, D.; Baxter, D. Levels of inorganic constituents in raw nuts and seeds on the Swedish market. *Sci. Total Environ.* **2008**, *392*, 290–304. [CrossRef]
29. Sharif, R.; Paul, R.; Bhattacharjya, D.; Ahmed, K. Physicochemical characters of oilseeds from selected mustard genotypes. *J. Bangladesh Agric. Univ.* **2017**, *15*, 27–40. [CrossRef]
30. Vera, R.; Junior, M.S.; Naves, R.V.; Souza, E.R.B.; Fernandes, E.P.; Caliari, M.; Leandro, W.M. Características químicas de amêndoas de barueiros (*Dipteryx alata* vog.) de ocorrência natural no cerrado do estado de Goiás, Brasil. *Rev. Br. Frutic.* **2009**, *31*, 112–118. [CrossRef]
31. Glew, R.H.; Glew, R.S.; Chuang, L.-T.; Huang, Y.-S.; Millson, M.; Constans, D.; Vanderjagt, D.J. Amino Acid, mineral and fatty acid content of pumpkin seeds (*Cucurbita* spp.) and *Cyperus esculentus* nuts in the Republic of Niger. *Plant Foods Hum. Nutr.* **2006**, *61*, 49–54. [CrossRef]
32. Pang, Y.; MacIntosh, D.L.; Ryan, P.B. A longitudinal investigation of aggregate oral intake of copper. *J. Nutr.* **2001**, *131*, 2171–2176. [CrossRef] [PubMed]

33. Scheiber, I.F.; Mercer, J.F.B.; Dringen, R. Metabolism and functions of copper in brain. *Prog. Neurobiol.* **2014**, *116*, 33–57. [CrossRef] [PubMed]
34. Hernández-Sánchez, A.; Tejada-González, P.; Arteta-Jiménez, M. Aluminium in parenteral nutrition: A systematic review. *Eur. J. Clin. Nutr.* **2013**, *67*, 230–238. [CrossRef] [PubMed]

© 2020 by the authors. Licensee MDPI, Basel, Switzerland. This article is an open access article distributed under the terms and conditions of the Creative Commons Attribution (CC BY) license (http://creativecommons.org/licenses/by/4.0/).

*Article*

# Rapid Identification and Visualization of Jowl Meat Adulteration in Pork Using Hyperspectral Imaging

**Hongzhe Jiang \*, Fengna Cheng and Minghong Shi**

College of Mechanical and Electronic Engineering, Nanjing Forestry University, Nanjing 210037, China; cfn1218@163.com (F.C.); mhshi@njfu.edu.cn (M.S.)
\* Correspondence: jianghongzhe@njfu.edu.cn

Received: 31 December 2019; Accepted: 4 February 2020; Published: 6 February 2020

**Abstract:** Minced pork jowl meat, also called the sticking-piece, is commonly used to be adulterated in minced pork, which influences the overall product quality and safety. In this study, hyperspectral imaging (HSI) methodology was proposed to identify and visualize this kind of meat adulteration. A total of 176 hyperspectral images were acquired from adulterated meat samples in the range of 0%–100% (w/w) at 10% increments using a visible and near-infrared (400–1000 nm) HSI system in reflectance mode. Mean spectra were extracted from the regions of interests (ROIs) and represented each sample accordingly. The performance comparison of established partial least square regression (PLSR) models showed that spectra pretreated by standard normal variate (SNV) performed best with $R_p^2$ = 0.9549 and residual predictive deviation (RPD) = 4.54. Furthermore, functional wavelengths related to adulteration identification were individually selected using methods of principal component (PC) loadings, two-dimensional correlation spectroscopy (2D-COS), and regression coefficients (RC). After that, the multispectral RC-PLSR model exhibited the most satisfactory results in prediction set that $R_p^2$ was 0.9063, RPD was 2.30, and the limit of detection (LOD) was 6.50%. Spatial distribution was visualized based on the preferred model, and adulteration levels were clearly discernible. Lastly, the visualization was further verified that prediction results well matched the known distribution in samples. Overall, HSI was tested to be a promising methodology for detecting and visualizing minced jowl meat in pork.

**Keywords:** hyperspectral imaging; jowl meat; minced pork; meat adulteration; visualization

---

## 1. Introduction

Meat always plays an important role in the constitution of human diets around the world. People consume meat mainly due to its rich nutritional contents of essential amino acid and vitamins [1]. With the continuous and rapid increase of meat consumption nowadays, meat quality and safety control has become the major priority [2]. Meat is quite susceptible to suffering adulteration, such as the known horsemeat scandal in 2013 that horsemeat was detected in beef. From the perspective of consumers, they highly desire reliable and clear information about the meat or meat products they purchase [3]. Therefore, regardless of deliberate, accidental, or economically motivated meat adulteration, rapid identification is always one of the main issues in further prevention.

Minced meat, also named ground meat, is one of the most popular meat types. It is versatile in that it can be the major ingredient of a variety of meat products including sausages, patties, hamburgers, and meatballs [4]. Especially in Chinese diet culture, traditional dumplings, steamed stuffed buns, or wontons fillings are also mainly composed of this important ingredient. However, the morphological structure of muscles is removed when meat is minced so that the occurrence of adulteration in minced meat is hardly recognized by visual analysis. Pork jowl meat is a kind of lymphatic meat, which contains thyroid, lipomas, and significant amounts of lymph nodes [5]. It is low-price, full of stench, and should be discarded after the animals are slaughtered [5]. In the present period, a relatively high rate of fraudulent phenomena occur in China involving minced jowl meat being substituted for or added into minced pork to be served as fillings in food. These incidents seriously pose health treats and violate rights for consumers. Meat adulteration is not routinely detected, thus, it is highly desirable to rapidly identify if the fillings are adulterated with jowl meat.

Several detection techniques including high-performance liquid chromatography [6], polymerase chain reaction [7], mass spectrometry [8], differential scanning calorimetry [9], and enzyme-linked immunosorbent assays [10] were demonstrated to be effective in detecting meat adulteration. However, these techniques need complex sample preparation and destructive operation, which are high-cost, time-consuming, and laborious. More recently, a considerable number of fast optical techniques have shown potential in detecting meat adulteration. Among them, UV-visible [11], near-infrared spectroscopy (NIRS) [12,13], mid-infrared spectroscopy [14], Raman spectroscopy [15], laser induced breakdown spectroscopy [16], and computer vision [17] have been successfully developed to identify various adulterants in meat. Typically, conventional NIRS could overcome the above-mentioned drawbacks of traditional techniques for its sensitive, rapid, reagent-free, and nondestructive nature, and it has great potential to adapt real-time monitoring application [18,19]. However, the main limitation is that the spectral single-point detection in preselected areas cannot well represent the whole heterogeneous meat sample. In this light, interest in hyperspectral imaging (HSI) is continuously growing. HSI integrates conventional spectroscopy and imaging to acquire both spectral and spatial information from an identical sample to provide the traceable chemical and physical qualities simultaneously [20]. It has received a lot of attention and interest in inspecting both raw and processed meat items [21,22].

With regard to the detection of meat adulteration using HSI, several examples have previously been reported, for instance, pork or beef adulterated with chicken [17], chicken adulterated with carrageenan [23], beef adulterated with pork [24,25], prawn adulterated with gelatin [26], lamb adulterated with pork [27], beef adulterated with horsemeat [28,29], lamb adulterated with duck [30], etc. All the above studies achieved good performance through a combination of HSI and chemometrics. However, to date, few studies focused on the adulteration of minced pork with minced jowl meat.

Therefore, the main aim of this study is to investigate the feasibility of visible and near-infrared (VNIR, 400–1000 nm) HSI for detecting minced jowl meat in pork. The specific objectives were (1) to establish partial least square regression (PLSR) models based on spectra extracted from hyperspectral images and (2) to identify effective wavelengths for developing multispectral models and visualizing the adulteration.

## 2. Materials and Methods

### 2.1. Sample Preparation

Pure pork meat (from *Longissimus dorsi* muscle) and adulterant of jowl meat from the homologous slaughtered porcine body were purchased from a local retail market in Nanjing, China and transported to our laboratory within 30 min. On the day of purchase, pork meat was first cut into small pieces and minced using a meat grinder (S2-A808, Joyoung Co. Ltd., Jinan, China) for 60 s. The grinder was carefully washed using detergent and hot water, rinsed with distilled water, wiped with paper towel, and totally dried before jowl meat mincing use. In the same way, jowl meat was subsequently minced to be used as the adulterant. Minced pork samples were adulterated by mixing the minced adulterant into pork in range of 10%–90% (w/w) at 10% increments. Additionally, pure pork (0% adulteration level) and pure jowl meat (100% adulteration level) samples were also prepared. Sixteen replicates were prepared for each adulteration level, and samples were individually weighted. They were thoroughly mixed together to obtain a roughly homogenous paste with a final weight constant at 50 g. Then, all the prepared samples were put into round disposable Petri dishes (9 cm in diameter × 1.4 cm deep) with flat surfaces for subsequent imaging procedure.

In total, 176 samples including 16 samples at each adulteration level (11 levels) were prepared. Among them, three quarters of the samples at each adulteration level (12 samples × 11 levels = 132 samples) were randomly selected to be assigned to calibration set, and the residual one quarter (4 samples × 11 levels = 44 samples) were used for purpose of independent prediction. Furthermore, in order to prove the creditability of visualization results, two more control samples with known distributed patterns were also prepared. As for control sample one, four fan-shaped parts at different individual adulteration levels of 100%, 80%, 40%, and 20% were included. With regard to control sample two, two semicircular areas at individual 0% and 50% adulteration levels were covered. After that, the hyperspectral images of all the prepared samples were subsequently captured.

### 2.2. Hyperspectral Image Acquisition and Calibration

All the prepared samples were scanned using a laboratory-based push-broom hyperspectral imaging system in reflectance mode. The system converted the visible and near-infrared (VNIR) spectral range of 400–1000 nm (284 spectral bands) to capture the hyperspectral images at room temperature (26 ± 1 °C). The system consisted of a computer (Lenovo Tianyi 510 Pro, Lenovo Group Ltd., Beijing, China) installed with data acquisition software (Spectral Image software, Isuzu Optics Corp., Taiwan, China), a spectrograph (ImSpectorV10E, Spectral Imaging Ltd., Oulu, Finland), a 12-bit charged couple device (CCD) camera (HScamera-VIS, Isuzu Optics Corp., Xinzhu, China) with a C-mount lens, an illumination unit of two 150-W tungsten-halogen lamps, and a translation stage (Specim Spectral Imaging Ltd., Oulu, Finland) powered by stepping motor (SC30021A, Zolix Instrument Co, Beijing, China). The spectral solution of the system was 2.8 nm, and the size of incident slit was 30 μm (width) × 14.2 mm (length).

After hyperspectral image acquisition, calibration was conducted using two reference images by the equation as follows:

$$R_c = (R_o - D)/(W - D) \times 100\% \tag{1}$$

where $R_c$ represents calibrated relative reflectance hyperspectral image, $R_o$ denotes acquired original hyperspectral image, D expresses dark reference hyperspectral image with about 0% reflectance, and W stands for white reference hyperspectral image with about 99.9% reflectance. This procedure was directly carried out through the images calibration function within data acquisition software.

### 2.3. Region of Interests (ROI) Identification and Spectral Extraction

To isolate pure meat portion in the acquired images, representative regions of interests (ROIs) were first taken based on the calibrated images. ROI was individually determined for each hyperspectral image by applying a corresponding binary mask which was first established using band math procedure. Initially, the low reflectance image at 450 nm was subtracted from the high reflectance image at 890 nm. ROI was formed within the resulting image by thresholding at a constant value of 0.2. The ROIs were completely isolated from backgrounds and edges of the Petri dishes. After that, the mask was built and applied to corresponding hyperspectral images. Then, average spectra were extracted to represent each corresponding sample. All the involved steps were conducted using functions of mask building and ROI generation in ENVI 5.3 (Research Systems Inc., Solutions, Boulder, CO, USA).

### 2.4. Spectral Pretreatments

In order to compare and obtain robust and reliable performance, spectral data pretreatment is necessary prior to the development of quantitative models. In this research, a series of pretreatments including normalization, standard normal variate (SNV), multiplicative signal correction (MSC), detrending (detrend), first-order derivative, and second-order derivative (1st and 2nd derivative) were applied in addition to non-preprocessed spectra. Multiplicative interferences of scatter in the spectra can be effectively removed using SNV approach. MSC is a method like SNV which is effectively used in multiplicative variations elimination. Derivatives were usually used to remove baseline offsets and separate overlapping absorption bands. In this research, derivatives were calculated using second-order polynomial with Savitzky–Golay smoothing by a moving window size of 15 points. Detrending was implemented combined with SNV to suppress the baseline shifting and curvilinearity. Normalization was utilized to present the spectral differences caused by slight optical path variations. The pretreatment or the combinations were all implemented in the Unscrambler X 10.1 (Camo Software Inc., Trondheim, Norway).

*2.5. Modeling Method*

PLSR is a reliable linear regression modeling method that has been widely employed in spectral analysis for quantitatively predicting agro-products' quality traits. This method is especially suitable in performing the situation where there is a linear relationship between attributes of the targets and variables. In this study, PLSR models were developed with the dataset in calibration set using a leave-one-out cross-validation (LOOCV) method to prevent data over-fitting. PLSR first projects the spectra onto a few orthogonal factors named latent variables (LVs) [31]. The optimal LVs were determined where lowest root mean square error value of cross-validation was achieved. The PLSR modeling procedure was carried out using the software MATLAB 2013b (MathWorks Inc., Natick, MA, USA) with the PLS toolbox.

*2.6. Wavelengths Selection Methods*

Principal component analysis (PCA) is an unsupervised exploratory technique that has been reported to be a powerful tool in dimensionality reduction and multivariate data visualization. The variances of the whole dataset are first explained by PCA, and only a few new orthogonal latent variables that maximize the data variance (called the principal components: PCs) are retained [32]. PCA helps to look for relationships among the samples, and samples in the same class will gather together in PC score plots. If a clear clustering of grouped samples is shown in the PC space by combining two or more PCs, corresponding PC loadings are effective to determine informative wavelengths. Then, pronounced peaks and valleys of the PC loadings are considered to contribute more to the spectral variations of samples with different adulteration levels. The PCA procedure was conducted using the MATLAB 2013b.

Two-dimensional correlation spectroscopy (2D-COS) is a commonly used analytical mathematical formalism. Recently, 2D-COS analysis is highly concerned with identifying a set of spectroscopic data including Raman, visible-infrared, and fluorescence under external perturbation [33]. In terms of generalized 2D-COS, the perturbation can be pressure, concentration, temperature, etc. [34]. To discuss the generated spectrum, synchronous spectrum is diagonal symmetry and there were several autopeaks located at diagonal line. The synchronous spectrum could be used to characterize differences of the spectral intensities at different wavelengths. If the spectral intensities changed sharply with levels at a certain wavelength, there will be a strong autopeak. Thus, in our study, the autopeaks were introduced and utilized as effective wavelengths for identifying the adulteration.

Regression coefficients (RC), which are always used in combination with PLSR modeling method, are also an effective wavelength selection approach. In PLSR models, peaks and valleys at certain wavelengths with dominated RC values indicate a high influence on the response (predicted results) [35]. These spectral variables would be more useful in PLSR modeling and should be chosen for further PLSR models simplification. In our study, wavelengths with high absolute RC values (above the cutoff threshold) from the optimal PLSR model were considered to contribute most in predicting adulteration levels and they were finally adopted.

## 2.7. Models Performance Assessment

In order to assess the performance of established models, the following criteria including coefficient of determination in calibration set ($R_c^2$), cross-validation ($R_{cv}^2$), and prediction sets ($R_p^2$), as well as root mean squared error in calibration (RMSEC), cross-validation (RMSECV), and prediction sets (RMSEP) were determined, respectively. Furthermore, residual predictive deviation (RPD) was also evaluated to assess the practical utility of prediction models. If the values of $R^2 \geq 0.70$ and RPD $\geq 2.00$, models were considered to be effective in detecting meat quality and safety [36]. A satisfactory model should perform with results of high values in $R_c^2$, $R_{cv}^2$, $R_p^2$, and RPD as well as low values in RMSEC, RMSECV, and RMSEP.

## 2.8. Distribution Maps of the Adulterant

Recognizing adulterate distribution in minced pork is helpful to rapidly observe general adulteration level or if the sample was adulterated. The generation of distribution map is a means of visualization. It is a special advantage of HSI that conventional imaging or NIRS could not achieve. The optimal simplified model can be applied back to predict values in each pixel in the multispectral images at selected wavelengths. After that, the distribution map which pieced all predicted values of pixels together is generated. Therefore, in this research, after the optimal simplified model was confirmed, a colorful image with a linear color scale used for visualizing adulteration levels was displayed. All these steps were performed using a homemade program developed in Matlab 2013b.

## 3. Results and Discussion

### 3.1. Spectral Profiles

The average reflectance spectra of all the adulterated samples (Figure 1a) and pure minced pork and jowl meat extracted from the corresponding hyperspectral images (Figure 1b) are presented in Figure 1. Different spectra showed similar patterns with certain differences in reflective intensity. As shown, minced pork samples had slightly higher reflective intensity than minced jowl meat samples in spectral region of 400–1000 nm. Although there were few overlaps between spectra of pork and jowl meat, it was still possible to observe certain spectral differences, especially at prominent peaks or valleys. The variations in spectral reflectance among pork and jowl meat were related to the differences in chemical composition, and this implies the adulteration would induce significant alterations to the pure pork samples in a way that can be detected using spectral information.

In the VNIR region, several downwards peaks (absorbance bands) could be observed for pork and jowl meat. The band centered at around 411 nm was associated with the Soret absorption band, which was due to a respiratory pigment of haemoglobin [37]. The valleys of 543 nm and 570 nm can be ascribed to the presentence of deoxymyoglobin and oxymyoglobin, respectively [38], which were responsible for color traits of meat. The 975 nm and 759 nm were attributed to the second and third overtone of O–H stretching mode of water, respectively [39,40]. As for a weak reflectance valley, the independent absorbance band of 842 nm could correspond to the C-H stretching mode of aliphatic compounds [41,42]. The comparison showed that the reflective differences at these bands indicated that minced jowl meat samples had more contents of myoglobin, fat, and water than pure pork. Thus, the identification of jowl meat adulteration in pork is preliminarily concluded to be feasible on the basis of its spectral characteristics difference.

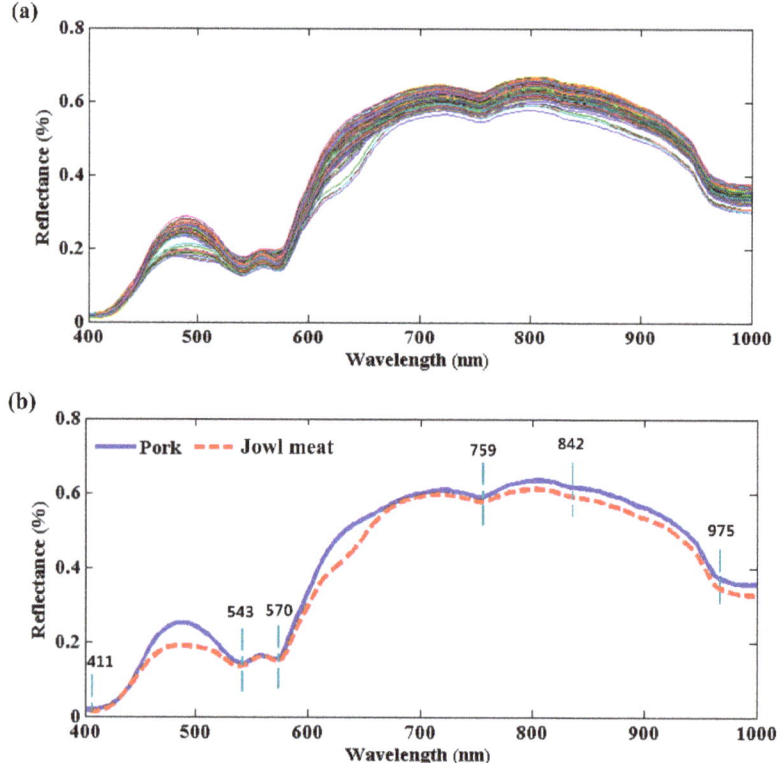

**Figure 1.** Spectral characteristics of prepared samples in the visible and near-infrared region. (a) Adulterated samples, (b) Mean raw spectra of pure pork and jowl meat.

*3.2. PLSR Models*

To determine and quantify the adulteration levels, PLSR models were developed based on raw or pretreated (normalization, SNV, MSC, SNV + Detrend, 1st derivative and 2nd derivative) spectra. A summary of the predictive results is listed in Table 1. As can be seen, spectral data with or without various pretreatments all showed good capability in predicting the adulteration levels. The overall $R^2$ values were higher than 0.94, RMSE values were lower than 10.2%, and RPD values were higher than 3.1. The small differences observed among RMSEC, RMSECV, and RMSEP values indicated that models were robust and reliable. Overall, HSI coupled with PLSR modeling method provided an innovative way to perform the instant and noncontact prediction for jowl meat adulterated in pork. With regard to a comparison of different pretreatments, the best PLSR modeling results were obtained based on SNV pretreated spectral data with performance of $R_p^2 = 0.9549$, RMSEP = 7.04%, and RPD = 4.54. Thus, the pretreatment of SNV was adopted to preprocess spectra in subsequent analysis of model simplification and visualization.

Table 1. Performance of partial least square regression (PLSR) models for predicting minced jowl meat adulterated in pork.

| Pretreatments | LVs | Calibration | | Cross-Validation | | Prediction | | |
|---|---|---|---|---|---|---|---|---|
| | | $R_c^2$ | RMSEC | $R_{cv}^2$ | RMSECV | $R_p^2$ | RMSEP | RPD |
| None | 12 | 0.9866 | 3.64% | 0.9779 | 4.71% | 0.9458 | 7.50% | 4.27 |
| Normalization | 14 | 0.9898 | 3.18% | 0.9821 | 4.24% | 0.9493 | 7.25% | 4.41 |
| SNV | 12 | 0.9864 | 3.68% | 0.9787 | 4.60% | 0.9549 | 7.04% | 4.54 |
| MSC | 13 | 0.9878 | 3.50% | 0.9801 | 4.45% | 0.9536 | 7.06% | 4.53 |
| SNV + Detrend | 12 | 0.9870 | 3.59% | 0.9797 | 4.51% | 0.9512 | 7.47% | 4.28 |
| 1st derivative | 13 | 0.9886 | 3.36% | 0.9815 | 4.30% | 0.9528 | 7.19% | 4.45 |
| 2nd derivative | 14 | 0.9896 | 3.23% | 0.9797 | 4.50% | 0.9425 | 10.18% | 3.14 |

Notes: PLSR: partial least squares regression; SNV: standard normal variate; MSC: multiplicative scatter correction; LVs: latent variables; $R_c^2$: coefficient of determination in calibration set; $R_{cv}^2$: coefficient of determination in cross-validation set; $R_p^2$: coefficient of determination in prediction set; RMSEC: root mean squared error for calibration set; RMSECV: root mean squared error for cross-validation set; RMSEP: root mean squared error for prediction set; RPD: residual predictive deviation.

### 3.3. Wavelengths Selection

#### 3.3.1. PCA Explanatory Analysis

Principal component analysis (PCA) is an efficient chemometric method, which provides the interpretation of variances among different data points in spectral analysis. In order to compare and highlight the spectral similarities and differences, PCA was first applied to the whole dataset of 176 spectra. The first three PCs which individually accounted for 80.37%, 9.50%, and 6.26% of the total variance were retained. The reason was that above 95% of the variation could be explained by the first three PCs. Moreover, through trial and error with different PC combinations, $PC_1$ and $PC_3$ were found to be useful in grouping samples into different adulteration levels. Then, the calculated PC scores of data points for different adulteration levels were utilized to create a 2D score plot (Figure 2). In general, data points in the same adulteration level tended to gather together and will be separated from others. The score plot of the combination of $PC_1$ vs. $PC_3$ is shown in Figure 2a. As the adulteration level continued to rise, corresponding samples tended to move along the positive directions of $PC_1$ axis and $PC_3$ axis. However, in low-level adulteration (less than 30%), data clusters were observed to be quite close and overlapped in a certain level. Tracing the root of the above observation, the main chemical composition of homologous pork and jowl meat was too similar so that no distinct separation was displayed if adulteration level was low. In addition, $PC_2$ seemed to mainly express the mutual information of samples with different adulteration levels, which was eliminated in the discrimination.

The PC loading lines of the two effective PCs were analyzed in detail and plotted in Figure 2b. Wavelengths at pronounced peaks and valleys were considered to carry important information in identifying the adulteration and should be selected. As a result, a total of nine wavelengths (440 nm, 491 nm, 545 nm, 560 nm, 570 nm, 632 nm, 686 nm, 752 nm, and 871 nm) were chosen. The valley at 440 nm could be attributed to deoxymyoglobin, the peak at 491 nm is associated with metmyoglobin, and 632 nm could be assigned to sulfmyoglobin [38]. The 676 nm is related to the presentation of redness, and the 871 nm band is relevant with the C-H vibration of hydrocarbons. The wavelengths selected by spectral PCA further confirmed the above results in spectral characteristics that pork and jowl meat were different in color presentation as well as water and hydrocarbon contents.

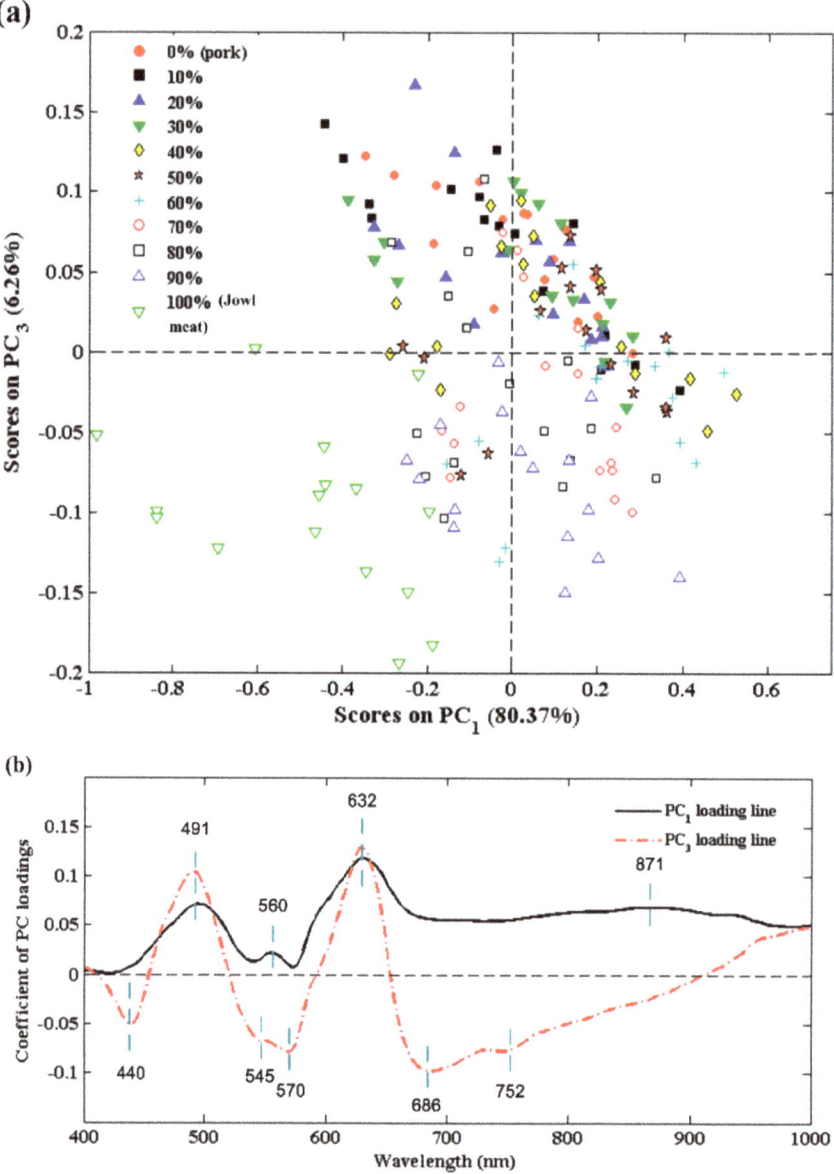

**Figure 2.** Analysis of effective PC scores and loadings. (**a**) PCA score plot of $PC_1$ vs. $PC_3$, (**b**) Wavelength selection on $PC_1$ and $PC_3$ loading lines.

## 3.3.2. Two-Dimensional Correction Spectroscopy

The 2D-COS analysis of the obtained 11 average spectra with adulteration levels from 0% to 100% is shown in Figure 3. There were two dominant autopeaks of 491 nm and 632 nm observed at the diagonal line in synchronous spectrum (Figure 3a). Another weak autopeak of 871 nm also occurred which could be clearly seen in the corresponding 3D stereo plot in Figure 3b. The presence of these suggested that intensity at these bands varied seriously with the adulteration levels. Therefore, these three wavelengths were effective in identifying the adulteration levels. It is worth mentioning that these three selected wavelengths were also included in the wavelengths selected by PC loadings. In terms of spectral variables, these three wavelengths are the most important in identifying the adulteration.

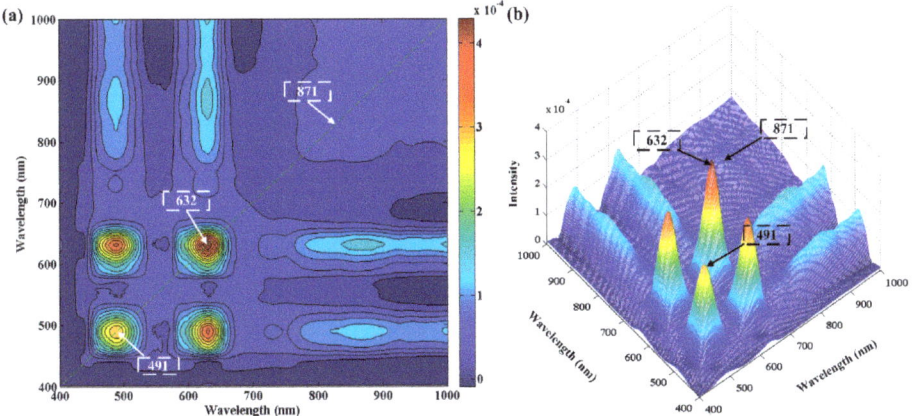

**Figure 3.** The 2D-COS spectrum of samples with various adulteration levels. (**a**) Synchronous contour map plot, (**b**) Corresponding synchronous 3D stereo plot.

### 3.3.3. Regression Coefficients

The regression coefficients (RC) curve from the preferred PLSR model based on SNV pretreated spectra is shown in Figure 4. The cut-off threshold was set to be 5, and only wavelengths at peaks and valleys with higher absolute coefficients than 5 were retained. Finally, a total of 10 (433 nm, 450 nm, 481 nm, 558 nm, 578 nm, 594 nm, 634 nm, 661 nm, 889 nm, and 948 nm) discontinuous wavelengths were deemed as the most effective wavelengths in PLSR models' development for quantitatively predicting jowl meat adulteration in minced pork. These wavelengths were different from the above ones but also reasonable due to the consideration of targeted prediction values.

Based on the above three wavelengths selection methods, the number of variables reduced significantly by at least 96.5% to at most 98.9%. The retained wavelengths could be utilized in developing a robust model and further a low-cost multispectral imaging system. Therefore, all the three groups of effective wavelengths could be the basis for comparison in developing the simplified PLSR models.

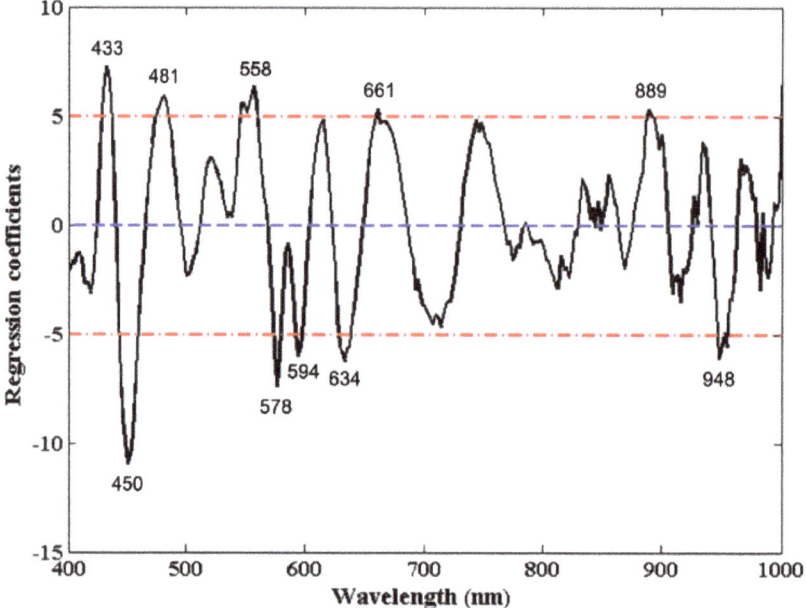

**Figure 4.** Regression coefficients of the optimal PLSR model based on full spectra.

*3.4. Multispectral Models Development*

In order to further eliminate the useless wavelengths and optimize the processing time in computing, simplified PLSR models based on selected wavelengths were established. Wavelengths selected by 2D-COS, PC loadings, and RC methods were individually set as inputs of the simplified PLSR models, and the overall results are displayed in Table 2. As can be seen, the results obtained using selected wavelengths slightly decreased compared with the full spectra. This phenomenon indicated that selected wavelengths were effective and the eliminated variables also contained little information in determining the adulteration. Compared with 2D-COS-PLSR and PC loadings-PLSR models, RC-PLSR model achieved the best performance with $R_p^2$ of 0.9063, RMSEP of 13.93%, and RPD of 2.30. It indicated that the 10 wavelengths selected by RC method were the most critical in identifying jowl meat adulteration in pork. On the contrary, the three wavelengths selected by 2D-COS performed not that well mainly because that they were not informative enough. As an extension, the additional six more wavelengths selected using PC loadings significantly improved the prediction accuracy by showing the $R_p^2$ of 0.7475, RMSEP of 17.31%, and RPD of 1.85. What is more, 2D-COS and PC loadings selected wavelengths using X-variables (spectra) only, while RC was based on PLSR model which decomposed both X-and Y-variables (adulteration levels) in the LVs calculation. RC built the optimal relationship between spectral data and adulteration levels compared with 2D-COS and PC loadings. Therefore, the multispectral RC-PLSR model was finally chosen for further visualization steps.

Table 2. Performance of simplified PLSR models based on wavelengths selected by three methods.

| Method | Number | LVs | Calibration | | Cross-Validation | | Prediction | | |
|---|---|---|---|---|---|---|---|---|---|
| | | | $R_c^2$ | RMSEC | $R_{cv}^2$ | RMSECV | $R_p^2$ | RMSEP | RPD |
| 2D-COS | 3 | 3 | 0.2283 | 27.78% | 0.1920 | 28.45% | 0.2720 | 27.45% | 1.17 |
| PC loadings | 9 | 6 | 0.8981 | 10.09% | 0.8344 | 10.80% | 0.7475 | 17.31% | 1.85 |
| RC | 10 | 9 | 0.9610 | 6.24% | 0.9520 | 6.93% | 0.9063 | 13.93% | 2.30 |

Notes: PLSR: partial least squares regression; 2D-COS: two-dimensional correction spectroscopy; PC: principal component; RC: regression coefficients; LVs: latent variables; $R_c^2$: coefficient of determination in calibration set; $R_{cv}^2$: coefficient of determination in cross-validation set; $R_p^2$: coefficient of determination in prediction set; RMSEC: root mean squared error for calibration set; RMSECV: root mean squared error for cross-validation set; RMSEP: root mean squared error for prediction set; RPD: residual predictive deviation.

The determination of the limit of detection (LOD) is an important step which investigated if the lowest adulteration concentration can be detected with the HSI methodology. The LOD was commonly calculated to evaluate the sensitivity of detection methods by the following equation [43].

$$LOD = 2\delta_b/S \tag{2}$$

where $\delta_b$ indicates the standard deviation (SD) of the background response and $S$ denotes the sensitivity by the ratio of the predicted adulteration levels to the reference values (namely the slope of the calibration line).

The performance of the preferred multispectral RC-PLSR model with error bar is illustrated in Figure 5. The results of this optimal model were visualized, and it could be seen that all the adulterated samples could be detected (predicted values were above 0%) whether in the calibration or prediction set. The LOD in independent prediction set calculated by the above Equation (2) was found to be 6.50%. However, the aim of meat adulteration is generally to make profit so that adulteration is always performed to be more than 10% [44]. Thus, the LOD in this research proved that it was competent in detecting jowl meat adulteration in pork by the HSI system.

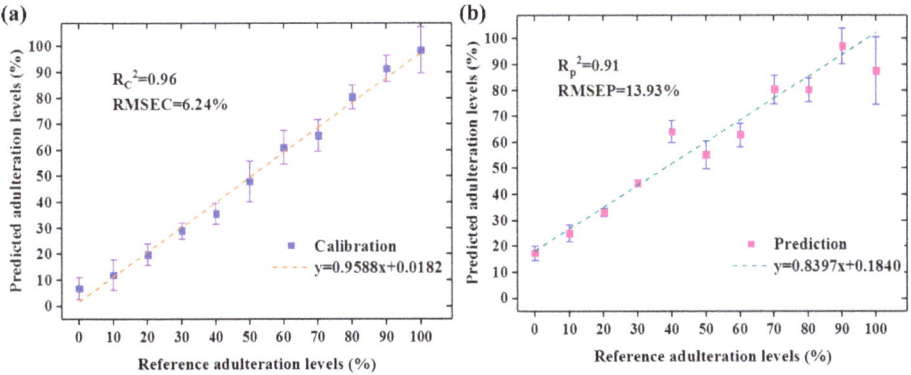

Figure 5. The performance of preferred multispectral RC-PLSR model with error bar in (a) Calibration set, and (b) Prediction set.

### 3.5. Visualization of the Adulteration Levels

As known, each sample was represented by the average spectra extracted from corresponding ROI, and the targeted adulteration level was only indicated by one value. However, there was abundant spatially distributed information in hyperspectral images. In this research, the adulteration level at each pixel in one hyperspectral image was predicted so that distribution was visualized for a quick view. The optimal RC-PLSR model was first applied to the multispectral images recombined at selected wavelengths. The false color images (left column) and corresponding predicted distribution maps (right column) of the samples with different adulteration levels in prediction set are shown in Figure 6. The false color image in Figure 6a was composited by setting the images at 700.9 nm, 545.4 nm, and 436.4 nm as R (red), G (green), and B (blue) channels using ENVI software. They were quite close to the true color image at the primary RGB colors' wavelengths (700 nm, 546.1 nm and 435.8 nm). As can be seen, actual adulteration level is difficult to be recognized by naked eyes in Figure 6a. Distribution maps expressed how the adulteration varied from sample to sample and even from pixel to pixel within one sample and were generated to be shown in Figure 6b. The linear color bar located in the right side from black to red indicated different adulteration levels from 0% to 100% accordingly. There was a clear tendency of color gradient with the increasing adulteration level so that adulteration was easily distinguishable in distribution maps.

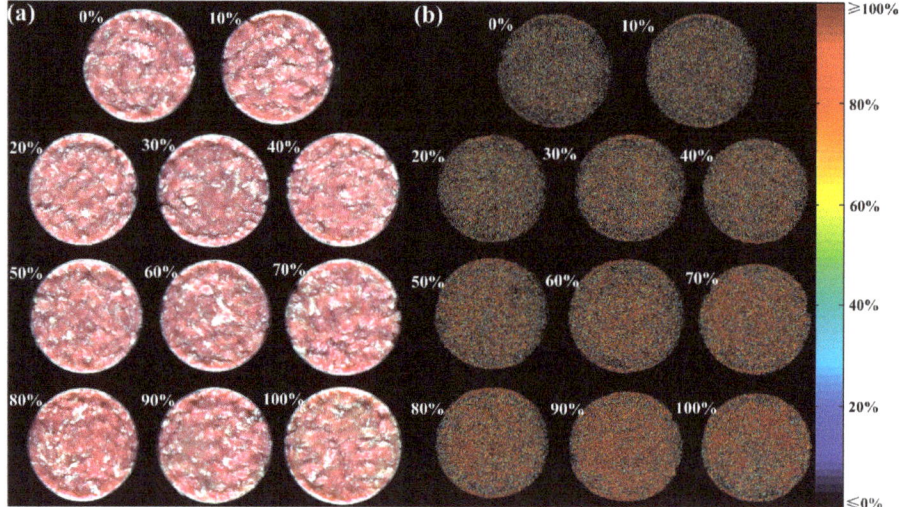

**Figure 6.** Distribution maps of jowl meat adulteration in pork at pixel level. (**a**) False color images, (**b**) Distribution maps.

In further steps, the reliability of visualization was verified through generating distribution maps for the two control samples. Two false color images were in the up row and corresponding distribution maps were displayed in the down row in Figure 7. For control sample one, the results of the distribution map showed that four fan-shaped areas were in different colors. The upper left area was mainly in red, and the lower left showed red and yellow. The upper right exhibited a wide range of colors while the lower right presented general blue. With regard to control sample two, the distribution map gave a display of half black and half yellow. All the prediction maps were generally consistent with the actual situation. The effectiveness of the established RC-PLSR model and visualization procedure was thus proved again.

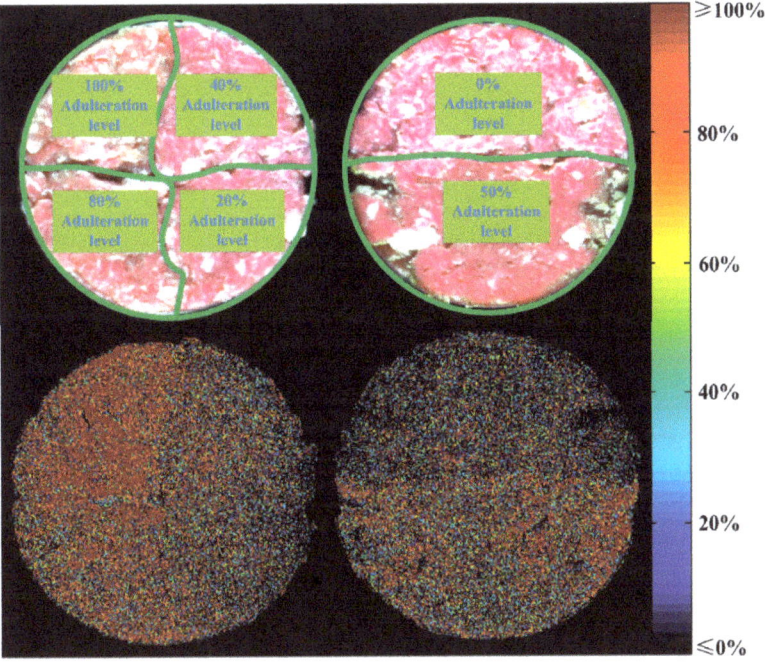

**Figure 7.** Visualization for control samples.

## 4. Conclusions

This study was motivated by the requirement for rapidly and nondestructively identifying one common adulterant of minced jowl meat in minced pork. Our attempts explored the application of HSI combined with chemometrics and wavelength selection algorithm to quantify and visualize the adulteration. In particular, simplified RC-PLSR model developed by 10 key wavelengths gave the best performance; LOD achieved 6.50%. The visualization of adulteration levels was successfully performed based on RC-PLSR model. Predicted colorful distribution maps were generated to make it convenient in observation. To test the validity of visualization, two known distributed samples were predicted and expected corresponding maps were displayed. The overall results suggested that HSI had the potential to identify minced jowl meat adulteration in pork without any prior physical or chemical analysis information. This technique could provide more detailed visualization information than conventional imaging and NIRS in identifying adulteration levels. However, more studies related to a large number of samples and different breeds in sampling should be conducted. In addition, further work will focus on identifying a variety of commonly used adulterates in pork based on a few wavelengths to achieve the goal for portable application.

**Author Contributions:** H.J. conceived and designed the experiments. H.J., F.C., and M.S. conducted experiments and analyzed the data. H.J. wrote the paper. All authors have read and agreed to the published version of the manuscript.

**Funding:** This research was funded by the Scientific Research Foundation for Advanced Talents of Nanjing Forestry University.

**Acknowledgments:** The authors would like to thank organizations or individuals who had contributed to this manuscript.

**Conflicts of Interest:** The authors declare no conflict of interest.

## References

1. De Smet, S.; Vossen, E. Meat: The balance between nutrition and health. A review. *Meat Sci.* **2016**, *120*, 145–156. [CrossRef] [PubMed]
2. Rahmati, S.; Julkapli, N.M.; Yehye, W.A.; Basirun, W.J. Identification of meat origin in food products—A review. *Food Control* **2016**, *68*, 379–390. [CrossRef]
3. Ballin, N.Z. Authentication of meat and meat products. *Meat Sci.* **2010**, *86*, 577–587. [CrossRef] [PubMed]
4. Alamprese, C.; Amigo, J.M.; Casiraghi, E.; Engelsen, S.B. Identification and quantification of turkey meat adulteration in fresh, frozen-thawed and cooked minced beef by FT-NIR spectroscopy and chemometrics. *Meat Sci.* **2016**, *121*, 175–181. [CrossRef] [PubMed]
5. Ding, H.; Shao, H.; Yang, Y. Discussion on the edible safety of pork jowl meat. *Shandong Food Ferment* **2013**, *03*, 56–58. (In Chinese)
6. Al-Rashood, K.A.; Abdel-Moety, E.M.; Rauf, A.; Abou-Shaaban, R.R.; Al-Khamis, K.I. Triacylglycerols-profiling by high performance liquid chromatography: A tool for detection of pork fat (lard) in processed foods. *J. Liq. Chromatogr. Relat. Technol.* **1995**, *18*, 2661–2673. [CrossRef]
7. He, H.; Hong, X.; Feng, Y.; Wang, Y.; Ying, J.; Liu, Q.; Qin, Y.; Zhou, X.; Wang, D. Application of quadruple multiplex PCR detection for beef, duck, mutton and pork in mixed meat. *J. Food Nutr. Res.* **2015**, *3*, 392–398. [CrossRef]
8. Black, C.; Chevallier, O.P.; Cooper, K.M.; Haughey, S.A.; Balog, J.; Takats, Z.; Elliott, C.T.; Cavin, C. Rapid detection and specific identification of offals within minced beef samples utilising ambient mass spectrometry. *Sci. Rep.* **2019**, *9*, 6295. [CrossRef]
9. Dahimi, O.; Rahim, A.A.; Abdulkarim, S.M.; Hassan, M.S.; Hashari, S.B.Z.; Mashitoh, A.S.; Saadi, S. Multivariate statistical analysis treatment of DSC thermal properties for animal fat adulteration. *Food Chem.* **2014**, *158*, 132–138. [CrossRef]
10. Macedo-Silva, A.; Barbosa, S.F.C.; Alkmin, M.D.G.A.; Vaz, A.J.; Shimokomaki, M.; Tenuta-Filho, A. Hamburger meat identification by dot-ELISA. *Meat Sci.* **2000**, *56*, 189–192. [CrossRef]
11. Alamprese, C.; Casale, M.; Sinelli, N.; Lanteri, S.; Casiraghi, E. Detection of minced beef adulteration with turkey meat by UV–Vis, NIR and MIR spectroscopy. *LWT Food Sci. Technol.* **2013**, *53*, 225–232. [CrossRef]
12. Pieszczek, L.; Czarnik-Matusewicz, H.; Daszykowski, M. Identification of ground meat species using near-infrared spectroscopy and class modeling techniques—Aspects of optimization and validation using a one-class classification model. *Meat Sci.* **2018**, *139*, 15–24. [CrossRef] [PubMed]
13. Morsy, N.; Sun, D.W. Robust linear and non-linear models of NIR spectroscopy for detection and quantification of adulterants in fresh and frozen-thawed minced beef. *Meat Sci.* **2013**, *93*, 292–302. [CrossRef] [PubMed]
14. Zhao, M.; Downey, G.; O'Donnell, C.P. Detection of adulteration in fresh and frozen beefburger products by beef offal using mid-infrared ATR spectroscopy and multivariate data analysis. *Meat Sci.* **2014**, *96*, 1003–1011. [CrossRef]
15. Zhao, M.; Downey, G.; O'Donnell, C.P. Dispersive Raman spectroscopy and multivariate data analysis to detect offal adulteration of thawed beefburgers. *J. Agric. Food Chem.* **2015**, *63*, 1433–1441. [CrossRef]
16. Velioglu, H.M.; Sezer, B.; Bilge, G.; Baytur, S.E.; Boyaci, I.H. Identification of offal adulteration in beef by laser induced breakdown spectroscopy (LIBS). *Meat Sci.* **2018**, *138*, 28–33. [CrossRef]
17. Nolasco-Perez, I.M.; Rocco, L.A.; Cruz-Tirado, J.P.; Pollonio, M.A.; Barbon, S.; Barbon, A.P.A.; Barbin, D.F. Comparison of rapid techniques for classification of ground meat. *Biosyst. Eng.* **2019**, *183*, 151–159. [CrossRef]
18. Jiang, X.; Zhao, T.; Liu, X.; Zhou, Y.; Shen, F.; Ju, X.; Liu, X.; Zhou, H. Study on Method for On-Line Identification of Wheat Mildew by Array Fiber Spectrometer. *Spectrosc. Spectr. Anal.* **2018**, *38*, 3729–3735.
19. Ge, Y.; Atefi, A.; Zhang, H.; Miao, C.; Ramamurthy, R.K.; Sigmon, B.; Yang, J.; Schnable, J.C. High-throughput analysis of leaf physiological and chemical traits with VIS–NIR–SWIR spectroscopy: A case study with a maize diversity panel. *Plant Methods* **2019**, *15*, 66. [CrossRef]
20. ElMasry, G.; Sun, D.W.; Allen, P. Chemical-free assessment and mapping of major constituents in beef using hyperspectral imaging. *J. Food Eng.* **2013**, *117*, 235–246. [CrossRef]
21. Barbin, D.F.; Sun, D.W.; Su, C. NIR hyperspectral imaging as non-destructive evaluation tool for the recognition of fresh and frozen–thawed porcine *longissimus dorsi* muscles. *Innov. Food Sci. Emerg. Technol.* **2013**, *18*, 226–236. [CrossRef]

22. Feng, C.H.; Makino, Y.; Oshita, S.; Martín, J.F.G. Hyperspectral imaging and multispectral imaging as the novel techniques for detecting defects in raw and processed meat products: Current state-of-the-art research advances. *Food Control* **2018**, *84*, 165–176. [CrossRef]
23. Zhang, Y.; Jiang, H.; Wang, W. Feasibility of the Detection of Carrageenan Adulteration in Chicken Meat Using Visible/Near-Infrared (Vis/NIR) Hyperspectral Imaging. *Appl. Sci.* **2019**, *9*, 3926. [CrossRef]
24. Ropodi, A.I.; Pavlidis, D.E.; Mohareb, F.; Panagou, E.Z.; Nychas, G.J. Multispectral image analysis approach to detect adulteration of beef and pork in raw meats. *Food Res. Int.* **2015**, *67*, 12–18. [CrossRef]
25. Kamruzzaman, M.; Makino, Y.; Oshita, S. Hyperspectral imaging in tandem with multivariate analysis and image processing for non-invasive detection and visualization of pork adulteration in minced beef. *Anal. Methods* **2015**, *7*, 7496–7502. [CrossRef]
26. Wu, D.; Shi, H.; He, Y.; Yu, X.; Bao, Y. Potential of hyperspectral imaging and multivariate analysis for rapid and non-invasive detection of gelatin adulteration in prawn. *J. Food Eng.* **2013**, *119*, 680–686. [CrossRef]
27. Kamruzzaman, M.; Sun, D.W.; ElMasry, G.; Allen, P. Fast detection and visualization of minced lamb meat adulteration using NIR hyperspectral imaging and multivariate image analysis. *Talanta* **2013**, *103*, 130–136. [CrossRef]
28. Ropodi, A.I.; Panagou, E.Z.; Nychas, G.J.E. Multispectral imaging (MSI): A promising method for the detection of minced beef adulteration with horsemeat. *Food Control* **2017**, *73*, 57–63. [CrossRef]
29. Kamruzzaman, M.; Makino, Y.; Oshita, S.; Liu, S. Assessment of visible near-infrared hyperspectral imaging as a tool for detection of horsemeat adulteration in minced beef. *Food Bioprocess Technol.* **2015**, *8*, 1054–1062. [CrossRef]
30. Zheng, X.; Li, Y.; Wei, W.; Peng, Y. Detection of adulteration with duck meat in minced lamb meat by using visible near-infrared hyperspectral imaging. *Meat Sci.* **2019**, *149*, 55–62. [CrossRef]
31. Menesatti, P.; Zanella, A.; D'Andrea, S.; Costa, C.; Paglia, G.; Pallottino, F. Supervised multivariate analysis of hyper-spectral NIR images to evaluate the starch index of apples. *Food Bioprocess Technol.* **2009**, *2*, 308–314. [CrossRef]
32. Wold, S.; Esbensen, K.; Geladi, P. Principal component analysis. *Chemom. Intell. Lab. Syst.* **1987**, *2*, 37–52. [CrossRef]
33. Dou, X.; Yuan, B.; Zhao, H.; Yin, G.; Ozaki, Y. Generalized two-dimensional correlation spectroscopy. *Sci. China Ser. B* **2004**, *47*, 257–266. [CrossRef]
34. Noda, I. Generalized two-dimensional correlation method applicable to infrared, Raman, and other types of spectroscopy. *Appl. Spectrosc.* **1993**, *47*, 1329–1336. [CrossRef]
35. He, H.J.; Wu, D.; Sun, D.W. Potential of hyperspectral imaging combined with chemometric analysis for assessing and visualising tenderness distribution in raw farmed salmon fillets. *J. Food Eng.* **2014**, *126*, 156–164. [CrossRef]
36. Kapper, C.; Klont, R.E.; Verdonk, J.M.A.J.; Urlings, H.A.P. Prediction of pork quality with near infrared spectroscopy (NIRS): 1. Feasibility and robustness of NIRS measurements at laboratory scale. *Meat Sci.* **2012**, *91*, 294–299. [CrossRef]
37. Mamani-Linares, L.W.; Gallo, C.; Alomar, D. Identification of cattle, llama and horse meat by near infrared reflectance or transflectance spectroscopy. *Meat Sci.* **2012**, *90*, 378–385. [CrossRef]
38. Liu, Y.L.; Chen, Y.R. Two-dimensional correlation spectroscopy study of visible and near-infrared spectral variations of chicken meats in cold storage. *Appl. Spectrosc.* **2000**, *54*, 1458–1470. [CrossRef]
39. Wu, D.; Sun, D.W. Application of visible and near infrared hyperspectral imaging for non-invasively measuring distribution of water-holding capacity in salmon flesh. *Talanta* **2013**, *116*, 266–276. [CrossRef]
40. Bowker, B.; Hawkins, S.; Zhuang, H. Measurement of water-holding capacity in raw and freeze-dried broiler breast meat with visible and near-infrared spectroscopy. *Poult. Sci.* **2014**, *93*, 1834–1841. [CrossRef]
41. Zhao, M.; O'Donnell, C.; Downey, G. Detection of offal adulteration in beefburgers using near infrared reflectance spectroscopy and multivariate modelling. *J. Near Infrared Spectrosc.* **2013**, *21*, 237. [CrossRef]
42. Barbin, D.F.; ElMasry, G.; Sun, D.W.; Allen, P. Non-destructive determination of chemical composition in intact and minced pork using near-infrared hyperspectral imaging. *Food Chem.* **2013**, *138*, 1162–1171. [CrossRef] [PubMed]

43. Gondal, M.A.; Seddigi, Z.S.; Nasr, M.M.; Gondal, B. Spectroscopic detection of health hazardous contaminants in lipstick using laser induced breakdown spectroscopy. *J. Hazard. Mater.* **2010**, *175*, 726–732. [CrossRef] [PubMed]
44. Bilge, G.; Velioglu, H.M.; Sezer, B.; Eseller, K.E.; Boyaci, I.H. Identification of meat species by using laser-induced breakdown spectroscopy. *Meat Sci.* **2016**, *119*, 118–122. [CrossRef] [PubMed]

© 2020 by the authors. Licensee MDPI, Basel, Switzerland. This article is an open access article distributed under the terms and conditions of the Creative Commons Attribution (CC BY) license (http://creativecommons.org/licenses/by/4.0/).

*Article*

# Validation of a Quantitative Proton Nuclear Magnetic Resonance Spectroscopic Screening Method for Coffee Quality and Authenticity (NMR Coffee Screener)

Alex O. Okaru [1], Andreas Scharinger [2], Tabata Rajcic de Rezende [2], Jan Teipel [2], Thomas Kuballa [2], Stephan G. Walch [2] and Dirk W. Lachenmeier [2,*]

1. Department of Pharmaceutical Chemistry, University of Nairobi, P.O. Box 19676-00202 Nairobi, Kenya; alex.okaru@gmail.com
2. Chemisches und Veterinäruntersuchungsamt (CVUA) Karlsruhe, Weissenburger Straße 3, 76187 Karlsruhe, Germany; andreas.scharinger@cvuaka.bwl.de (A.S.); tabata.rajcicderezende@cvuaka.bwl.de (T.R.d.R.); jan.teipel@cvuaka.bwl.de (J.T.); thomas.kuballa@cvuaka.bwl.de (T.K.); stephan.walch@cvuaka.bwl.de (S.G.W.)
* Correspondence: lachenmeier@web.de; Tel.: +49-721-926-5434

Received: 29 November 2019; Accepted: 1 January 2020; Published: 4 January 2020

**Abstract:** Monitoring coffee quality as a means of detecting and preventing economically motivated fraud is an important aspect of international commerce today. Therefore, there is a compelling need for rapid high throughput validated analytical techniques such as quantitative proton nuclear magnetic resonance (NMR) spectroscopy for screening and authenticity testing. For this reason, we sought to validate an $^1$H NMR spectroscopic method for the routine screening of coffee for quality and authenticity. A factorial experimental design was used to investigate the influence of the NMR device, extraction time, and nature of coffee on the content of caffeine, 16-*O*-methylcafestol (OMC), kahweol, furfuryl alcohol, and 5-hydroxymethylfurfural (HMF) in coffee. The method was successfully validated for specificity, selectivity, sensitivity, and linearity of detector response. The proposed method produced satisfactory precision for all analytes in roasted coffee, except for kahweol in canephora (robusta) coffee. The proposed validated method may be used for routine screening of roasted coffee for quality and authenticity control (i.e., arabica/robusta discrimination), as its applicability was demonstrated during the recent OPSON VIII Europol-Interpol operation on coffee fraud control.

**Keywords:** caffeine; 16-*O*-methylcafestol; kahweol; furfuryl alcohol; tetramethylsilane (TMS); magnetic resonance spectroscopy; validation studies

---

## 1. Introduction

Coffee remains a popular beverage worldwide and is typically obtained from the two species, namely *Coffea canephora* (robusta) and *Coffea arabica* [1–3]. *Coffea arabica* fetches a higher price in the market owing to its perceived superior organoleptic properties and higher production costs compared to *Coffea canephora* [4]. Consequently, beverage fraud involving complete or partial substitution of arabica with robusta coffee cannot be overruled. On the other hand, the diterpenes cafestol, 16-*O*-methylcafestol (OMC), and kahweol found in the lipid fraction of coffee serve as potential markers for the differentiation of *C. canephora* and *C. arabica*. Cafestol is found in both *C. canephora* and *C. arabica* while OMC is specific only to *C. canephora* [5–7]. Kahweol, although present in both types of coffee, is found in significantly higher amounts in *C. arabica*. These differences in the diterpene constituents enable the distinction between the coffees and also enable detection of beverage fraud involving substitution of C. arabica with the cheaper C. canephora beans using OMC as a marker [8].

A number of analytical techniques such as high performance liquid chromatography (HPLC) [9], gas chromatography with flame ionization detection [10], gas chromatography-mass spectrometry [11,12], proton transfer mass spectrometry [13], nuclear magnetic resonance (NMR) spectroscopy [14], isotope-ratio mass spectrometry [15], near-infrared spectroscopy [16,17], electronic nose [18], flame atomic absorption spectrometry [19], and attenuated total reflectance Fourier transform infrared spectroscopy [16], among other techniques, has been reported in the literature for the quantitative determination of coffee constituents and screening of coffee for adulteration.

Nuclear magnetic resonance spectroscopy in combination with chemometrics has been applied either for routine quality control and/or detection of potentially harmful substances in beverages such as alcohol [20], fruit juices [21,22] and coffee [8,23]. NMR spectroscopy may be applied for the quantification of caffeine, OMC, kahweol, 5-hydroxymethylfurfural (HMF) and furfuryl alcohol in coffee [24–26]. For decaffeinated coffee, NMR spectroscopy may be used to determine the residual quantities of caffeine, which would typically be less than 1 g/kg. Furfuryl alcohol and HMF may be used as indicators of the degree of coffee roasting [24]. However, furfuryl alcohol is also of public health significance and therefore may require monitoring using NMR. The International Agency for Research on Cancer (IARC) classifies furfuryl alcohol into Group 2B (possibly carcinogenic) [27]. NMR also offers the advantages of cost-effectiveness especially for screening. Additionally, NMR provides reproducible quantitative data [28,29] and generates unique chemical fingerprints that may be useful for authenticity testing [30,31]. Similar to other analytical techniques, reliable results may only be obtained by use of validated methods. Based on previously published method development and optimization work [14,24–26], the aim of this study was to validate the quantitative NMR spectroscopic method for screening coffee for both quality and authenticity.

## 2. Materials and Methods

### 2.1. Chemicals

Reagents and standard compounds were of analytical or HPLC grade. The five analytes caffeine, HMF, OMC, kahweol and furfuryl alcohol were purchased from Sigma-Aldrich (Steinheim, Germany). Deuterated chloroform-$d_1$ (≥99.8% atom % D) and internal reference standard tetramethylsilane (TMS) were obtained from Roth (Karlsruhe, Germany).

Reference Standards and Preparation of Working Standard Solutions

Primary stock solutions of caffeine, HMF, OMC, kahweol, and furfuryl alcohol were prepared in deuterated chloroform solution with 1% TMS ($CDCl_3$ + TMS). Individual stock solutions were prepared by separately dissolving 5 mg of caffeine, HMF, kahweol and furfuryl alcohol each in 5 mL $CDCl_3$ + TMS. For preparation of OMC stock solution, 10.9 mg of OMC powder was dissolved in 10.9 mL $CDCl_3$ + TMS. Working solutions were obtained by carrying out a 1:2 dilution. The stock solutions were kept in the freezer until use. The guidance concentration and defined working ranges for the working standards are given in Table 1. A control solution was prepared by dissolving 25.02 mg 1,2,4-tetrachloro-3-nitrobenzene with $CDCl_3$ to 5 mL.

Table 1. Working standards used.

| Substance | Guidance Value [a] (mg/kg) | Defined Working Ranges According to Experience (mg/kg) |
|---|---|---|
| OMC | <50 for arabica | 7.5–7500 |
| Caffeine | <1000 for decaf | 7.5–7500 |
| Kahweol | <300 for robusta | 7.5–7500 |
| Furfuryl alcohol | - | 7.5–7500 |
| HMF | - | 7.5–7500 |

[a] Guidance for OMC and kahweol based on own experience in analyzing coffee samples. Guidance for caffeine in decaf coffee is the limit in German national coffee regulation ("Kaffeeverordnung") [32].

## 2.2. Methodology

### 2.2.1. Samples and Sample Preparation

Coffee samples (commercial products from local supermarket in Karlsruhe, Germany) for analysis were prepared by weighing 200 mg of ground beans before being dissolved in 1.5 mL of $CDCl_3$ + TMS. The samples were shaken for 10 min or 20 min at 350 rpm on the shaking machine. The solutions were then membrane filtered (0.45 µm) before 600 µL of the filtrate was transferred to an NMR tube followed by analysis.

### 2.2.2. NMR Analysis

Two 400 MHz (9.4 T) field strength spectrometers were used to acquire proton NMR spectra: an AVANCE 400 Ultra Shield with a 5 mm PASEI 1H/D-13C Z-GRD probe, and an Ascend 400 with a BBI 400S1 H-BB-D-05 Z (each from Bruker, Rheinstetten, Germany). All samples were measured in 5 mm sample tubes (NMR tube DEU-Quant 5 mm, 7 inch) (Deutero, Kastellaun, Germany). The spectra were automatically acquired at 300.0 K under the control of Sample Track and ICON-NMR (Bruker BioSpin, Rheinstetten, Germany). Detailed information about measurement methodology is available in [26].

A waiting time of 5 min for temperature equilibration was used for every measurement. The NMR spectra were acquired using a Bruker pulse program (zg30) with 64 scans (NS) and 2 prior dummy scans (DS) with a relaxation delay (D1) of 30 s and an acquisition time of 7.97 s. The time domain was set to 131072 data points with a spectral width of 20.5503 ppm (8223.68 Hz) for UltraShield and 20.5617 ppm (8223.69 Hz) for Ascend. The size of the real spectrum (SI) was 262144. The receiver gain was set to 45.2. All spectra were recorded with the basopt mode. The acquisition parameters were constant for all spectra for pulse length–based concentration determination (PULCON, see Wider & Dreier [33]) measurement according to Lachenmeier et al. [26]. The free induction decay (FID) was multiplied with an exponential window function to achieve a line broadening of 0.30 Hz. The spectra were automatically phased and baseline-corrected (default settings) using TopSpin version 3.2 and 3.5 (Bruker Biospin, Rheinstetten, Germany).

### 2.2.3. Experimental Design

A factorial experimental design was adopted for the validation studies. For this purpose, six matrix calibration series, each consisting of two blanks and ten samples with increasing amounts namely 1, 5, 10, 25, 50, 100, 250, 500, 750, and 1000 mg/L of analyte were prepared. A factorial design was employed for the investigation of the influence of the three experimental factors, NMR spectrometer type, coffee type and shaking time (see Table 2). Each measurement series corresponds to a different combination of factor characteristics (see Supplementary Table S1 for full design).

**Table 2.** Factorial experimental design used.

| Array | Factor 1: NMR Device | Factor 2: Coffee Type | Factor 3: Shaking Time (min) |
|---|---|---|---|
| 1 | Ultrashield/Ascend | 100% arabica decaffeinated | 20 |
| 2 | Ultrashield/Ascend | 100% robusta | 20 |
| 3 | Ultrashield/Ascend | Green coffee | 20 |
| 4 | Ultrashield/Ascend | 100% arabica decaffeinated | 10 |
| 5 | Ultrashield/Ascend | 100% robusta | 10 |
| 6 | Ultrashield/Ascend | Green coffee | 10 |

2.2.4. Preparation of Working and Test Solutions

Stock solutions (1000 and 5000 mg/L) of each of the analytes comprising caffeine, HMF, OMC, kahweol, and furfuryl alcohol were used to prepare 1, 5, 10, 25, 50, 75, 125, 250, 500, 750, and 1000 mg/L working solutions. Additionally, two blanks were made for each of the measurement series. Separate test solutions were prepared for the three matrices (100% arabica decaffeinated, 100% robusta coffee and green coffee). The dilution matrix to achieve the desired concentration is shown in Table 3.

**Table 3.** Dilution matrix.

| Desired Calibration Concentration (mg/L) | Dilution of Stock Solution (1000 mg/L) | Desired Final Concentration (mg/L) | Volume per Stock Solution (µL) | Volume of all Analytes (µL) | Volume of $CDCl_3$ (µL) |
|---|---|---|---|---|---|
| 0 (Blank, 2×) | 0 | 0 | - | - | 1500 |
| 1 | 1:1000 | 1 | 1.5 | 7.5 | 1492.5 |
| 5 | 1:200 | 5 | 7.5 | 37.5 | 1462.5 |
| 10 | 1:100 | 10 | 15 | 75 | 1425 |
| 25 | 1:40 | 25 | 37.5 | 187.5 | 1312.5 |
| 50 | 1:20 | 50 | 75 | 375 | 1125 |
| 100 | 1:10 | 100 | 150 | 750 | 750 |
| 250 | 1:20 | 250 | 75 | 375 | 1125 |
| 500 | 1:10 | 500 | 150 | 750 | 750 |
| 750 | 1:6.66 | 750 | 225 | 1125 | 375 |
| 1000 | 1:5 | 1000 | 300 | 1500 | - |

Therefore, for the six matrix calibration series, a total of 72 test solutions was prepared. However, since all samples were run in two instruments (Ultrashield/Ascend), 144 measurement results were obtained (or 120 without the blank values).

2.3. Validation Studies

Three different coffee matrices spanning the broadest possible spectrum of different coffee constituents were used during validation. These consisted of decaffeinated coffee (decaf. arabica, matrix 1), robusta coffee (matrix 2), and raw coffee (green coffee, matrix 3). For the preparation of the spiked matrix samples, each pure substance was weighed before being dissolved in $CDCl_3$ and TMS solution (usually in 5–10 mL). Subsequently, the test samples were spiked in the specified concentration range (see Supplementary Table S1). The control solution was also run after a series of measurements in order to ascertain that analyses were properly performed so that test results obtained could be considered reliable.

2.3.1. Selectivity

The selectivity of each analyte was established by measurement of all analytes mixed in a solution without matrix. To achieve this, 100 µL of each of the five analytes were pipetted into a NMR tube followed by addition of 500 µL $CDCl_3$ before NMR analysis (desired concentration 500 mg/L).

Furthermore, all analytes in a solution were mixed with 100% arabica decaffeinated coffee in order to check possible matrix disturbances. This was achieved by pipetting 150 µL of each of the five analytes into an NMR tube followed by the addition of 750 µL CDCl$_3$, before adding 200 mg of coffee sample (desired concentration 100 mg/L). The solution was shaken for 20 min at 350 rpm on the shaker, then membrane filtered and used directly for NMR measurement. For comparison, an NMR spectrum of a coffee sample (without analytes) was also acquired. The coffee sample was prepared by dissolving 200 mg coffee sample in 1.5 mL CDCl$_3$, shaken for 20 min at 350 rpm on the shaker, membrane filtered, and used for NMR analysis.

### 2.3.2. Detection and Quantification Limits

In order to determine the detection limit, 9 spiking levels of different concentrations were added to decaffeinated arabica coffee (Matrix 1), processed, measured and evaluated. The detection and determination limits were determined in the lower working range according to the German norm DIN 32645 [34].

### 2.3.3. Precision and Recovery

For the determination of the measurement uncertainties and the recoveries, 9 spiking levels at different concentrations were added to the 3 matrices and processed. The measurement uncertainty was determined with the aid of ANOVA (all settings at default) using Design Expert Software V.7.0 (Stat-Ease Inc., Minneapolis, MN, USA).

## 2.4. Data Analysis and Quality Control

Peak areas in the 1D-proton NMR spectra were evaluated with the help of a compiled MatLab script. The peak areas were determined using a line fitting algorithm. Quantification was performed using the eretic factor, which was previously determined using a quant reference (for details see [26]). At the end of each measurement series, the control solution was measured as a safeguard. The assignment of the signal patterns and the determination of the exact position of the signals were performed by the analysis of a 2D-JRES-NMR spectrum. Note: due to the restricted solubility of caffeine in CDCl$_3$, an empirical factor of 6 for recalculation has to be used, as determined based on HPLC measurements using the German reference procedure [35].

## 2.5. Method Performance

The method was assessed for performance by calculating the standard deviation of the intra-laboratory reproducibility, recovery, robustness, limits of measurements, and the total uncertainty of the measurements as a function of concentration.

## 3. Results and Discussion

### 3.1. Validation

#### 3.1.1. Specificity and Selectivity

The use of working reference standards enabled accurate assignment of chemical shifts. The chemical shifts of the analytes in the different matrices are shown in Table 4. A representative spectrum of an authentic sample including magnifications of target resonances is provided in Figure 1. OMC presented a slight offset in the integration range that led to a too high integral due to matrix interferences (especially fatty acids in the field region higher to OMC). However, this problem was circumvented by integrating the range next to OMC (3.04–3.10 ppm), which is similarly affected by the same matrix interference, and subtracting it from the sum of the integral of OMC (also see an illustration of the problem in Figure 2).

Foods **2020**, 9, 47

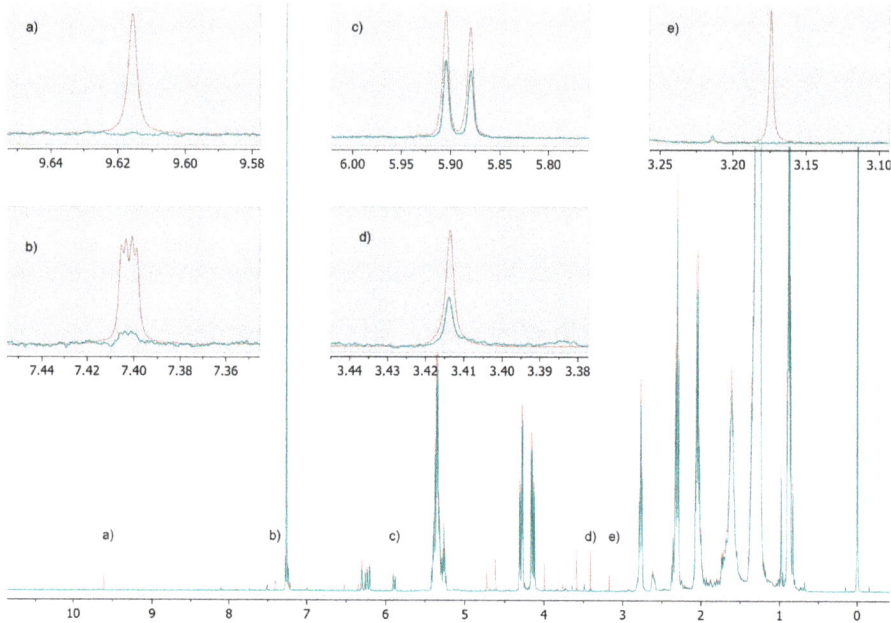

**Figure 1.** Representative 1H NMR spectrum of an authentic coffee sample showing target resonances in magnification (HMF (**a**), furfuryl alcohol (**b**), kahweol (**c**), caffeine (**d**), OMC (**e**)).

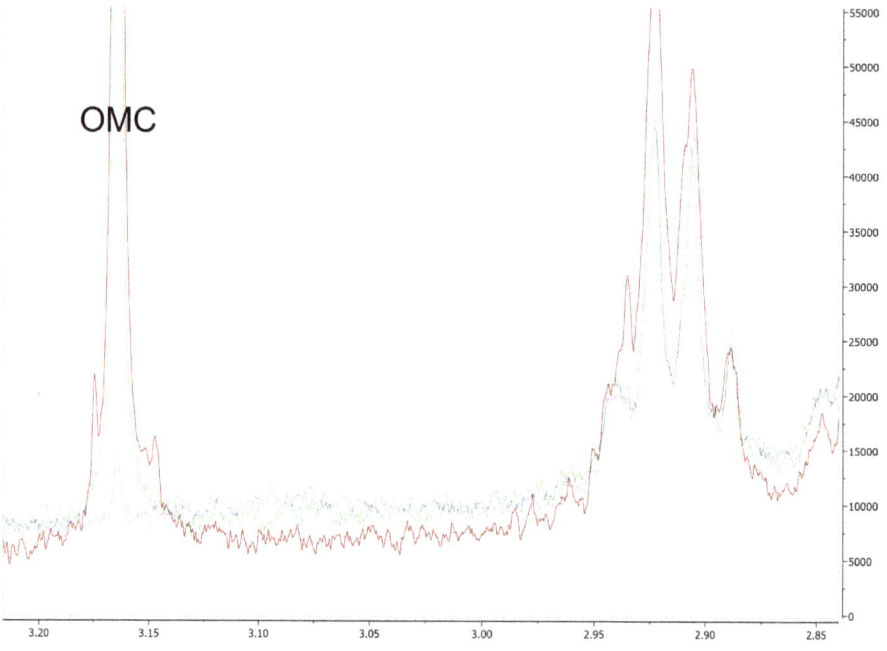

**Figure 2.** NMR range besides the OMC signal showing matrix interferences from resonances of fatty acid signals around 2.9 ppm. Over-quantification can be avoided by subtracting the noise range (3.04 ppm–3.10 ppm).

**Table 4.** Characteristic signals of the constituents in coffee and their ranges.

| Analyte | Integration Range (ppm) |
| --- | --- |
| OMC | 3.185–3.125 |
| Caffeine | 3.44–3.38 |
| Kahweol | 5.925–5.85 |
| Furfuryl alcohol | 7.411–7.39 |
| HMF | 9.69–9.67 |

3.1.2. Analytical Limits

The detection and quantification limits are given in Table 5 together with the concentration ranges. The limits of measurements were adjudged fit for purpose.

**Table 5.** Limits of detection and quantification of analytes determined.

| Analyte | Detection Limit (mg/kg) | Determination Limit (mg/kg) | Concentration Range for Determination of Limit (mg/kg) |
| --- | --- | --- | --- |
| OMC | 2.5 | 7.4 | 7.5–187.5 |
| Caffeine | 15.7 | 43.1 | 7.5–187.5 |
| Kahweol | 186.0 | 501.4 | 187.5–1875.0 |
| Furfuryl alcohol | 11.6 | 39.4 | 7.5–75 |
| HMF | 6.3 | 22.9 | 7.5–75 |

3.1.3. Precision

The recoveries of the different analytes in various matrices are shown in Table 6. Although, the recoveries in a majority of the matrices used were within limits, green coffee gave poor unsatisfactory recoveries for caffeine, OMC and kahweol. Moreover, the recovery of kahweol from robusta coffee was out of specifications too (see Table 6).

**Table 6.** Recovery of coffee constituents from different matrices.

| Matrix | Recovery (%) | | | | |
| --- | --- | --- | --- | --- | --- |
| | Caffeine | OMC | Kahweol | Furfuryl Alcohol | HMF |
| Decaf. arabica | 101 | 97 | 95 | 97 | 102 |
| Robusta | 102 | 101 | 74 * | 99 | 101 |
| Green coffee | 137 * | 54 * | 188 * | 93 | 107 |

* Outside of specification. Specification: 90–110%.

The coefficient of variation (CV) was used as criterion for evaluating the precision of the proposed NMR method. The acceptance criterion for precision was a CV of less than 15% (internal quality standard of the authors' laboratory). Apart from kahweol, the precision of all the other analytes was found being within the limits of acceptance in all matrices. The analytes, caffeine, OMC, furfuryl alcohol and HMF present in roasted coffee (arabica and robusta) can therefore be determined with sufficient precision and accuracy by using the proposed NMR method. However, kahweol may not be quantified with adequate precision in robusta due to its low content in this matrix. In addition to the out-of-specification recoveries, the precision of all the analytes for green coffee were unsuitable (Table 7). Further work, potentially by improving the extraction, appears to be necessary for green coffee.

Table 7. Precision of coffee constituents.

| Matrix | Precision (CV) | | | | |
|---|---|---|---|---|---|
| | Caffeine | OMC | Kahweol | Furfuryl Alcohol | HMF |
| Decaf. arabica | 8.1 | 6.5 | 22.2 * | 6.1 | 8.3 |
| Robusta | 7.4 | 7.8 | 32.7 * | 5.8 | 6.9 |
| Green coffee | 104 * | 188 * | 570 * | 25 * | 27 * |

* Outside of specification. Specification: <15%. CV—coefficient of variation.

### 3.1.4. Linearity of Detector Response

Linearity was established in the concentration ranges (working range) listed in Table 8. The linearity was determined in matrix 1. Since the coefficients of determination ($R^2$) were all >0.99 over the concentration ranges examined, the method may be considered to be fit-for-purpose.

Table 8. Linear concentration range of the coffee analytes (also see Supplementary Figure S1).

| Analyte | Linearity (mg/kg) | Coefficient of Determination ($R^2$) |
|---|---|---|
| OMC | 7.5–5625 | 1.0000 |
| Caffeine | 7.5–5625 | 1.0000 |
| Kahweol | 7.5–5625 | 0.9949 |
| Furfuryl alcohol | 7.5–5625 | 1.0000 |
| HMF | 7.5–5625 | 0.9997 |

### 3.1.5. Effect of Matrix

ANOVA revealed that the models are significant for all analytes and can be evaluated. For all analytes it was shown that the instrument used (i.e., NMR spectrometer type) has no significant influence on the analytical results. The measurements can thus be performed on both spectrometers. Similarly, the extraction time had no significant influence. If the results are viewed manually, the extraction time of 20 min seems adequate, but not statistically significant, to achieve better results, and was thus defined as a setting.

However, the influence of variety of coffee was found to be statistically significant especially with green coffee, which had a significantly greater dispersion. Roasting was found to have no influence on the determinations since similar recoveries were obtained for the analytes. The method can therefore only be considered successfully validated for the determination of OMC, caffeine, kahweol, furfuryl alcohol, and HMF in samples of roasted coffee. Measurements of green coffee shall be considered as indicative only.

### 3.1.6. Applicability

The method was applied to 797 samples since 2016. Suspicious samples, i.e., cases of potential food fraud (arabica samples with OMC > 50 mg/kg) were in all cases positively confirmed using the German norm procedure based on HPLC [35]. Furthermore, the applicability of the method was proven during the recent OPSON VIII Europol-Interpol operation [36,37], in which more than 150 roasted coffee samples were analyzed using the validated NMR procedure within the two-week operation period (see, e.g., [36]). In this sample, three cases of substantial admixture of robusta into coffee claimed as 100% arabica could be determined.

## 4. Conclusions

The proposed NMR spectroscopic method gave satisfactory validation results for specificity, selectivity and linearity. All analytes examined gave satisfactory recoveries except caffeine, OMC and kahweol in green coffee and kahweol in robusta coffee (due to its very low content in this matrix). The analytical limits were found to be adequate for routine NMR measurements for the analytes.

Importantly, the proton NMR spectroscopic method was found to be suitable for unambiguously coffee screening and authenticity testing. Additionally, the method may be adopted for the routine quantitation of furfuryl alcohol in coffee in analytical laboratories.

**Supplementary Materials:** The following are available online at http://www.mdpi.com/2304-8158/9/1/47/s1, Table S1: Raw results of method validation for coffee using a factorial experimental design; Figure S1: Linearity determination of target compounds.

**Author Contributions:** Conceptualization, D.W.L.; methodology, A.S., J.T., T.K. and D.W.L.; formal analysis, J.T., T.R.d.R. and A.S.; resources, S.G.W. and D.W.L.; data curation, A.S., J.T., T.R.d.R., T.K.; writing—original draft preparation, A.O.O. and D.W.L.; writing—review and editing, D.W.L. and A.O.O.; supervision, D.W.L. and S.G.W. All authors have read and agreed to the published version of the manuscript.

**Funding:** This research received no external funding.

**Acknowledgments:** Cornelia Ritter is thanked for excellent technical assistance.

**Conflicts of Interest:** The authors declare no conflict of interest.

## References

1. Nuhu, A.A. Bioactive micronutrients in coffee: Recent analytical approaches for characterization and quantification. *ISRN Nutr.* **2014**, *2014*, 384230. [CrossRef] [PubMed]
2. Samoggia, A.; Riedel, B. Consumers' perceptions of coffee health benefits and motives for coffee consumption and purchasing. *Nutrients* **2019**, *11*, 635. [CrossRef] [PubMed]
3. Higdon, J.V.; Frei, B. Coffee and health: A review of recent human research. *Crit. Rev. Food Sci. Nutr.* **2006**, *46*, 101–123. [CrossRef] [PubMed]
4. Hameed, A.; Hussain, S.A.; Suleria, H.A.R. "Coffee bean-related" agroecological factors affecting the coffee. In *Co-Evolution of Secondary Metabolites*; Merillon, J.M., Ramawat, K., Eds.; Reference Series in Phytochemistry; Springer: Cham, Switzerland, 2018. [CrossRef]
5. Scharnhop, H.; Winterhalter, P. Isolation of coffee diterpenes by means of high-speed countercurrent chromatography. *J. Food Compos. Anal.* **2009**, *22*, 233–237. [CrossRef]
6. Speer, K.; Kölling-Speer, I. The lipid fraction of the coffee bean. *Braz. J. Plant Physiol.* **2006**, *18*, 201–216. [CrossRef]
7. Finotello, C.; Forzato, C.; Gasparini, A.; Mammi, S.; Navarini, L.; Schievano, E. NMR quantification of 16-O-methylcafestol and kahweol in *Coffea canephora var. robusta* beans from different geographical origins. *Food Control* **2017**, *75*, 62–69. [CrossRef]
8. Schievano, E.; Finotello, C.; De Angelis, E.; Mammi, S.; Navarini, L. Rapid authentication of coffee blends and quantification of 16-O-methylcafestol in roasted coffee beans by nuclear magnetic resonance. *J. Agric. Food Chem.* **2014**, *62*, 12309–12314. [CrossRef]
9. Smrke, S.; Kroslakova, I.; Gloess, A.N.; Yeretzian, C. Differentiation of degrees of ripeness of Catuai and Tipica green coffee by chromatographical and statistical techniques. *Food Chem.* **2015**, *174*, 637–642. [CrossRef]
10. Jumhawan, U.; Putri, S.P.; Bamba, T.; Fukusaki, E. Application of gas chromatography/flame ionization detector-based metabolite fingerprinting for authentication of Asian palm civet coffee (Kopi Luwak). *J. Biosci. Bioeng.* **2015**, *120*, 555–561. [CrossRef]
11. Jumhawan, U.; Putri, S.P.; Marwani, E.; Bamba, T.; Fukusaki, E. Selection of discriminant markers for authentication of asian palm civet coffee (Kopi Luwak): A metabolomics approach. *J. Agric. Food Chem.* **2013**, *61*, 7994–8001. [CrossRef]
12. Mancha Agresti, P.D.C.; Franca, A.S.; Oliveira, L.S.; Augusti, R. Discrimination between defective and non-defective Brazilian coffee beans by their volatile profile. *Food Chem.* **2008**, *106*, 787–796. [CrossRef]
13. Özdestan, Ö.; van Ruth, S.M.; Alewijn, M.; Koot, A.; Romano, A.; Cappellin, L.; Biasioli, F. Differentiation of specialty coffees by proton transfer reaction-mass spectrometry. *Food Res. Int.* **2013**, *53*, 433–439. [CrossRef]
14. Monakhova, Y.B.; Ruge, W.; Kuballa, T.; Ilse, M.; Winkelmann, O.; Diehl, B.; Thomas, F.; Lachenmeier, D.W. Rapid approach to identify the presence of Arabica and Robusta species in coffee using $^1$H NMR spectroscopy. *Food Chem.* **2015**, *182*, 178–184. [CrossRef] [PubMed]

15. Rodrigues, C.; Brunner, M.; Steiman, S.; Bowen, G.J.; Nogueira, J.M.F.; Gautz, L.; Prohaska, T.; Máguas, C. Isotopes as tracers of the Hawaiian coffee-producing regions. *J. Agric. Food Chem.* **2011**, *59*, 10239–10246. [CrossRef] [PubMed]
16. Medina, J.; Caro Rodríguez, D.; Arana, V.A.; Bernal, A.; Esseiva, P.; Wist, J. Comparison of attenuated total reflectance mid-infrared, near infrared, and $^1$H-nuclear magnetic resonance spectroscopies for the determination of coffee's geographical origin. *Int. J. Anal. Chem.* **2017**, *2017*, 7210463. [CrossRef] [PubMed]
17. Esteban-Díez, I.; González-Sáiz, J.M.; Sáenz-González, C.; Pizarro, C. Coffee varietal differentiation based on near infrared spectroscopy. *Talanta* **2007**, *71*, 221–229. [CrossRef] [PubMed]
18. Dong, W.; Zhao, J.; Hu, R.; Dong, Y.; Tan, L. Differentiation of Chinese robusta coffees according to species, using a combined electronic nose and tongue, with the aid of chemometrics. *Food Chem.* **2017**, *229*, 743–751. [CrossRef] [PubMed]
19. Grembecka, M.; Malinowska, E.; Szefer, P. Differentiation of market coffee and its infusions in view of their mineral composition. *Sci. Total Environ.* **2007**, *383*, 59–69. [CrossRef]
20. Monakhova, Y.B.; Schäfer, H.; Humpfer, E.; Spraul, M.; Kuballa, T.; Lachenmeier, D.W. Application of automated eightfold suppression of water and ethanol signals in 1H NMR to provide sensitivity for analyzing alcoholic beverages. *Magn. Reson. Chem.* **2011**, *49*, 734–739. [CrossRef]
21. Monakhova, Y.B.; Schütz, B.; Schäfer, H.; Spraul, M.; Kuballa, T.; Hahn, H.; Lachenmeier, D.W. Validation studies for multicomponent quantitative NMR analysis: The example of apple fruit juice. *Accredit. Qual. Assur.* **2014**, *19*, 17–29. [CrossRef]
22. Spraul, M.; Schütz, B.; Rinke, P.; Koswig, S.; Humpfer, E.; Schäfer, H.; Mörtter, M.; Fang, F.; Marx, U.C.; Minoja, A. NMR-based multi parametric quality control of fruit juices: SGF profiling. *Nutrients* **2009**, *1*, 148–155. [CrossRef] [PubMed]
23. Defernez, M.; Wren, E.; Watson, A.D.; Gunning, Y.; Colquhoun, I.J.; Le Gall, G.; Williamson, D.; Kemsley, E.K. Low-field $^1$H NMR spectroscopy for distinguishing between arabica and robusta ground roast coffees. *Food Chem.* **2017**, *216*, 106–113. [CrossRef] [PubMed]
24. Lachenmeier, D.W.; Schwarz, S.; Teipel, J.; Hegmanns, M.; Kuballa, T.; Walch, S.G.; Breitling-Utzmann, C.M. Potential antagonistic effects of acrylamide mitigation during coffee roasting on furfuryl alcohol, furan and 5-hydroxymethylfurfural. *Toxics* **2019**, *7*, 1. [CrossRef]
25. Okaru, A.O.; Lachenmeier, D.W. The food and beverage occurrence of furfuryl alcohol and myrcene—Two emerging potential human carcinogens? *Toxics* **2017**, *5*, 9. [CrossRef] [PubMed]
26. Lachenmeier, D.W.; Teipel, J.; Scharinger, A.; Kuballa, T.; Walch, S.G.; Grosch, F.; Bunzel, M.; Okaru, A.O.; Schwarz, S. Fully automated identification of coffee species and simultaneous quantification of furfuryl alcohol using NMR spectroscopy. *J. AOAC Int.* **2020**, in press. [CrossRef]
27. Grosse, Y.; Loomis, D.; Guyton, K.Z.; El Ghissassi, F.; Bouvard, V.; Benbrahim-Tallaa, L.; Mattock, H.; Straif, K.; International Agency for Research on Cancer Monograph Working Group. Some chemicals that cause tumours of the urinary tract in rodents. *Lancet Oncol.* **2017**, *18*, 1003–1004. [CrossRef]
28. Malz, F.; Jancke, H. Validation of quantitative NMR. *J. Pharm. Biomed. Anal.* **2005**, *38*, 813–823. [CrossRef]
29. Bharti, S.K.; Roy, R. Quantitative $^1$H NMR spectroscopy. *Trends Anal. Chem.* **2012**, *35*, 5–26. [CrossRef]
30. Lachenmeier, D.; Schönberger, T.; Ehni, S.; Schütz, B.; Spraul, M. A discussion about the potentials and pitfalls of quantitative nuclear magnetic resonance (qNMR) spectroscopy in food science and beyond. In Proceedings of the XIII International Conference on the Applications of Magnetic Resonance in Food Science, Karlsruhe, Germany, 7–10 June 2016; pp. 77–85. [CrossRef]
31. Humpfer, E.; Schütz, B.; Fang, F.; Cannet, C.; Mörtter, M.; Schäfer, H.; Spraul, M. Food NMR optimized for industrial use-an NMR platform concept. In *Magnetic Resonance in Food Science: Defining Food by Magnetic Resonance*; Royal Society of Chemistry: Cambridge, UK, 2015; pp. 77–83. [CrossRef]
32. Verordnung über Kaffee, Kaffee- und Zichorien-Extrakte vom 15. November 2001 (BGBl. I S. 3107), die zuletzt durch Artikel 6 der Verordnung vom 5. Juli 2017 (BGBl. I S. 2272) geändert worden ist. Available online: https://www.gesetze-im-internet.de/kaffeev_2001/BJNR310700001.html (accessed on 18 December 2019). (In German).
33. Wider, G.; Dreier, L. Measuring protein concentrations by NMR spectroscopy. *J. Am. Chem. Soc.* **2006**, *128*, 2571–2576. [CrossRef]
34. *DIN 32645:2008-11. Chemical Analysis—Decision Limit, Detection Limit and Determination Limit under Repeatability Conditions— Terms, Methods, Evaluation*; DIN e. V.: Berlin, Germany, 2008.

35. *DIN 10779:2011-03. Analysis of Coffee and Coffee Products—Determination of 16-O-Methyl Cafestol Content of Roasted Coffee—HPLC-Method*; DIN e. V.: Berlin, Germany, 2008.
36. Lachenmeier, D.W. CVUA Karlsruhe unterstützt die Europol/INTERPOL-Operation OPSON VIII zur Aufklärung von Lebensmittelbetrug bei Kaffee. 2019. Available online: https://www.ua-bw.de/pub/beitrag.asp?subid=2&Thema_ID=2&ID=2985&lang=DE&Pdf=No (accessed on 18 December 2019). (In German)
37. Europol. Over €100 Million Worth of Fake Food and Drinks Seized in Latest Europol-Interpol Operation. 2019. Available online: https://www.europol.europa.eu/newsroom/news/over-%E2%82%AC100-million-worth-of-fake-food-and-drinks-seized-in-latest-europol-interpol-operation (accessed on 18 December 2019).

© 2020 by the authors. Licensee MDPI, Basel, Switzerland. This article is an open access article distributed under the terms and conditions of the Creative Commons Attribution (CC BY) license (http://creativecommons.org/licenses/by/4.0/).

*Article*

# Front-Face Fluorescence Spectroscopy and Chemometrics for Quality Control of Cold-Pressed Rapeseed Oil During Storage

Ewa Sikorska [1,\*], Krzysztof Wójcicki [1], Wojciech Kozak [1], Anna Gliszczyńska-Świgło [1], Igor Khmelinskii [2], Tomasz Górecki [3], Francesco Caponio [4], Vito M. Paradiso [4], Carmine Summo [4] and Antonella Pasqualone [4]

1. Institute of Quality Science, Poznań University of Economics and Business, al. Niepodległości 10, 61-875 Poznań, Poland; krzysztof.wojcicki@ue.poznan.pl (K.W.); w.kozak@ue.poznan.pl (W.K.); anna.gliszczynska-swiglo@ue.poznan.pl (A.G.-Ś.)
2. Department of Chemistry and Pharmacy and Center of Electronics, Optoelectronics and Telecommunications, Faculty of Science and Technology, University of the Algarve, FCT, DQF and CEOT, Campus de Gambelas, 8005-139 Faro, Portugal; ikhmelin@ualg.pt
3. Faculty of Mathematics and Computer Science, Adam Mickiewicz University, Uniwersytetu Poznańskiego 4, 61-614 Poznań, Poland; tomasz.gorecki@amu.edu.pl
4. Food Science and Technology Unit, Department of Soil, Plant and Food Sciences, University of Bari, via Amendola 165/a, I-70126 Bari, Italy; francesco.caponio@uniba.it (F.C.); vitomichele.paradiso@uniba.it (V.M.P.); carmine.summo@uniba.it (C.S.); antonella.pasqualone@uniba.it (A.P.)
* Correspondence: ewa.sikorska@ue.poznan.pl

Received: 11 November 2019; Accepted: 5 December 2019; Published: 10 December 2019

**Abstract:** The aim of this study was to test the usability of fluorescence spectroscopy to evaluate the stability of cold-pressed rapeseed oil during storage. Freshly-pressed rapeseed oil was stored in colorless and green glass bottles exposed to light, and in darkness for a period of 6 months. The quality deterioration of oils was evaluated on the basis of several chemical parameters (peroxide value, acid value, $K_{232}$ and $K_{270}$, polar compounds, tocopherols, carotenoids, pheophytins, oxygen concentration) and fluorescence. Parallel factor analysis (PARAFAC) of oil excitation-emission matrices revealed the presence of four fluorophores that showed different evolution throughout the storage period. The fluorescence study provided direct information about tocopherol and pheophytin degradation and revealed formation of a new fluorescent product. Principal component analysis (PCA) performed on analytical and fluorescence data showed that oxidation was more advanced in samples exposed to light due to the photo-induced processes; only a very minor effect of the bottle color was observed. Multiple linear regression (MLR) and partial least squares regression (PLSR) on the PARAFAC scores revealed a quantitative relationship between fluorescence and some of the chemical parameters.

**Keywords:** fluorescence; rapeseed oil; multiway analysis; parallel factor analysis (PARAFAC); multivariate regression

## 1. Introduction

The popularity of cold-pressed vegetable oils has increased in recent years due to the tendency of consumers to choose less-processed, healthy foods [1]. The most popular is olive oil with its well established health-promoting properties. Rapeseed oil is a valuable alternative to olive oil. It has a favorable composition of unsaturated fatty acids, from the nutritional point of view; the relation between linoleic (ω-6) and α-linolenic (ω-3) acids in this oil is 2:1. Moreover, it is a rich source of tocopherols and phytosterols [2–4].

The quality of oil is affected by oxidation processes that may lead to the loss of nutritional value, deterioration of sensory properties, and formation of toxic products. Oil oxidation proceeds via auto- or photo-oxidation processes, which involve, respectively, triplet or singlet oxygen. Autooxidation of oils involves radical forms of acylglycerols. In photosensitized oxidation, chlorophyll acts as a photosensitizer for the formation of singlet oxygen $^1O_2$, which reacts directly with double bonds of fatty acids. The hydroperoxides formed during oxidation decompose to produce off-flavor compounds [5]. The rate of oxidation depends on a variety of factors, including chemical composition of oil, temperature, exposure to light, and presence of oxygen.

Rapeseed oil has lower oxidative stability as compared to olive oil [6,7]. Several studies have been performed to assess quality degradation of rapeseed oil during photo- and autoxidation and storage [8–14]. It was found that the overall quality during storage was influenced by the type of container material (plastics, glass), storage conditions (light, temperature, oxygen availability), and time [13].

Glass is the most popular material used for bottling oils. It is one of the most inert and easy to clean materials. However, colorless glass transmits radiation in the entire visible range, leading to photooxidation of oils and reduction of their shelf-life. Colored glass bottles are used to prevent or limit photooxidation. Green glass bottles are often used to protect oil from light in the 300–500 nm wavelength range [15].

A variety of methods have been used to study oil oxidation and deterioration during shelf-life [15–17]. In addition to traditional techniques, spectroscopy coupled with chemometrics was found useful in assessing various aspects of oil quality. Among the spectroscopic techniques, fluorescence provides two main advantages—sensitivity and selectivity, enabling evaluation of minor oil components. Several studies have reported successful applications of oil autofluorescence in the analysis of oil oxidation [18–27]. Various measurement techniques were used in these studies, including measurements of excitation spectra [18], emission spectra [22,25,26], synchronous fluorescence spectra [19–21], and excitation-emission matrices [23,24,27]. This last technique provides the most comprehensive characteristics of fluorescent systems, because excitation-emission matrices are determined by both absorption and emission properties of a sample. Most of the fluorescence studies of oils were conducted using conventional spectrofluorimeters. Recently, a fluorometer with Light Emitting Diode (LED) flashlight excitation and a smartphone was developed for edible oil authentication using a hue-based fluorescence method [28].

The aim of the present study was to evaluate quality changes occurring in cold-pressed rapeseed oil during storage for 6 months under different conditions by means of chemical parameters and fluorescence spectroscopy. Parallel factor analysis (PARAFAC) was applied to decompose the excitation-emission matrices. Principal component analysis (PCA) allowed visualization of the relations between differently stored oil samples and the variables describing their properties. The relations between chemical changes and oil fluorescence were studied quantitatively by means of multiple linear regression (MLR) and partial least squares regression (PLSR) performed on PARAFAC scores.

## 2. Materials and Methods

### 2.1. Samples and Storage Conditions

Rapeseed oil freshly pressed in a local oil mill was used in the study. A volume of 200 cm$^3$ of oil was placed into 250 cm$^3$ glass bottles, either colorless or green. The transmission spectra of colorless and green bottles are presented in Figure 1.

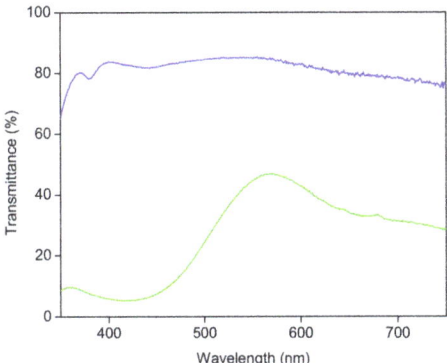

**Figure 1.** The transmission spectra of colorless (blue line) and green (green line) glass bottles.

The bottling procedure was the same as used in the oil mills. The samples were stored for a period of 6 months (from July to December). The storage conditions were similar to those used in the distribution and marketing of oils. Oil samples were divided into three groups and kept in appropriate conditions: (1) oils in colorless glass bottles, stored without light, marked RD; (2) oils in colorless glass bottles exposed to light, marked RL; (3) oils in green glass bottles, exposed to light, marked RG. The samples were exposed to diffused daylight and additionally for 12 h per day to artificial fluorescent cool white, 4000 K light (Osram light bulb) and illuminance of about 500 lx. Storage under diffused light simulated the conditions of a supermarket shelf. Storage in the dark simulated the warehouse storage. All bottles were stored at ambient room temperatures (18–25 °C). The oils samples were analyzed at the time of packaging and periodically during the 6-month period. Each month, one bottle corresponding to each of the storage conditions was opened and analyzed.

A total number of 18 samples was used for analysis; a single sample stored for 4 months in a colorless bottle exposed to light was excluded from analysis due to a leaking bottle cap.

### 2.2. Determination of Chemical Parameters

Oxygen concentrations in the bottle headspace and in the oil were determined using a commercially available OxySense 325i system. This instrument enabled non-invasive measurement of the oxygen content through the bottle wall, based on the effect of oxygen on the fluorescence lifetime of optically-excited Tris (4,7-diphenyl-1,10-phenanthroline) ruthenium chloride complex immobilized in a highly stable polymer in the form of an OxyDot oxygen indicator. The system consists of an oxygen concentration analyzer and an EasAlign pen with a built-in temperature sensor and OxyDot indicator. The OxyDot indicators were placed on the internal bottle wall in the headspace and in the oil phase using a transparent adhesive. The calibration of the sensors was performed using pure nitrogen (5.0 purity) and ambient air as standards.

Peroxide values expressed in milliequivalents of active oxygen/kg (meq $O_2$/kg of oil) were determined by the iodometric method, according to International Organization for Standardization (ISO) 3960:2007 standard [29].

Acid values were determined according to the standard ISO method 660:2009 [30].

The conjugated dienes and trienes were measured for samples diluted in n-hexane, using a Genesys UV-VIS spectrophotometer (Milton Roy, Houston, USA) and expressed as specific absorption coefficients at 232 ($K_{232}$) and 270 nm ($K_{270}$), according to the standard ISO method 3656:2011 [31].

Polar compounds (PCs) were separated from the oil samples by silica gel column chromatography, according to the Association of Official Analytical Chemists International (AOAC) method no. 982.27 [32]. After elution of the non-polar components with 150 mL of petroleum ether-diethyl ether (87:13, $v/v$), the PCs were recovered with 150 mL of diethyl ether. After removing the diethyl ether, the PCs were recovered in tetrahydrofuran (THF) and subjected to high performance size exclusion chromatography (HPSEC) analysis to determine the single classes of substances constituting them. The chromatographic system consisted of a series 200 Perkin–Elmer (Beaconsfield, UK) pump, a 50 µL injector loop (Perkin–Elmer, Beaconsfield, UK), a PL-gel guard column (Polymer Laboratories, Church Stretton, UK) of 5 cm length × 7.5 mm inner diameter, and a series of two PL-gel columns (Polymer Laboratories, Church Stretton, UK) of 30 cm length × 7.5 mm inner diameter each. The columns were packed with highly cross-linked styrene divinylbenzene copolymers with a particle diameter of 5 µm and pore diameters of 500 Å. The detector was a differential refractometer (series 200A, Perkin–Elmer, Beaconsfield, UK). The elution solvent used was THF for high performance liquid chromatography (HPLC) at a flow rate of 1.0 mL/min. The identification and quantification of individual peaks was carried out, as described in previous paper [33]. Three replicates were analyzed per sample.

Total tocopherol content was determined by using a HPLC method. The details of the method are described in Reference [34]. The analysis was carried out using a Waters 600 high-performance liquid chromatograph. Chromatographic separation was achieved at room temperature, using a Symmetry C18 column (150 mm × 3.9 mm, 5 µm), fitted with a µBondapak C18 cartridge guard column (all from Waters, Milford, MA, USA). A mobile phase composed of 50% acetonitrile and 50% methanol was used with a flow rate of 1 mL/min. Samples of oil were weighed (0.0800 g) and dissolved in 1 mL of 2-propanol. Vortex-mixed samples were directly injected into the HPLC column without any additional sample treatment. The injection volume was 20 µL. The eluate was detected using a Waters 474 scanning fluorescence detector, set for emission at 325 and excitation at 295 nm. The emission slit width was 10 nm, fluorometer gain 100, and attenuation 1. Tocopherols were identified by comparing their retention times with those of the corresponding standards. Additionally, a Waters 996 photodiode array detector was used to identify the compounds by their absorption spectra.

Carotenoids were determined using the spectrophotometric method. The concentration of total carotenoids was determined by measuring the absorption of oil dissolved in n-hexane at 450 nm. The molar absorption coefficient at 450 nm for carotenoids: $\varepsilon = 138730$ [L/cm × mol] was used for the calculation of carotenoid concentration [35]. Genesys 2 UV-VIS spectrophotometer (Milton Roy, Houston, USA) was used in these measurements.

Pheophytins content in fresh samples (expressed as mg of pheophytin as per kg of oil) was determined according to official methods and recommended practices of the American Oil Chemists' Society [36]. The absorbance of the oil sample was read at 630, 670, and 710 nm. In samples exposed to light, low concentration of pheophytins prevented their determination by means of the absorption method. Therefore, relative pheophytins content in all of the samples was determined based on the fluorescence intensity measured at $\lambda_{exc} = 670$ nm and $\lambda_{em} = 680$ nm, performed using a Fluorolog 3-11 spectrofluorometer (Spex–Jobin Yvon, Palaiseau, France). The relative pheophytins content in all samples studied was expressed as a percentage for further analysis.

*2.3. Fluorescence Measurements*

The fluorescence spectra were obtained using a Fluorolog 3-11 spectrofluorometer (Spex–Jobin Yvon). The excitation-emission matrices (EEMs) were collected by measuring the emission spectra in the 260–700 nm range with the excitation in the 250–500 nm range, with a 10 nm interval. The excitation and emission slit widths were 3 nm. The acquisition step and the integration time were maintained at 1 nm and 0.1 s, respectively. A reference photodiode detector was used to compensate for the source intensity fluctuations. The individual spectra were corrected for the wavelength-dependent response of the system. Front-face geometry was used for undiluted samples in a 10 mm fused-silica cuvette.

*2.4. Data Analysis*

The analysis of covariance (ANCOVA) was used to compare the chemical data for the samples stored in different conditions. Pearson correlation coefficients were calculated to test the correlations between the individual analytical parameters.

A parallel factor analysis (PARAFAC) was used to decompose excitation-emission matrices into contributions from individual fluorescent components. In PARAFAC, each component consists of one score vector and two loading vectors. The loading matrices contain the spectral excitation and emission profiles of fluorescent components, and the score matrix contains information about the relative contribution of each component to every sample EEMs included into the model [37].

A three-way array of data with the dimensions of 18 × 431 × 26 (number of samples × number of emission wavelengths × number of excitation wavelengths) was used for the PARAFAC. Rayleigh signals in EEMs were removed by replacing them with missing values. Non-negativity constraints on all three modes were applied. The optimal number of components in PARAFAC models was determined based on the core consistency diagnostic (CORCONDIA) and analysis of explained variance [37].

Principal component analysis (PCA) was performed on the X matrix, which contains chemical parameters and fluorescence score data obtained from the PARAFAC. The X data were scaled prior to analysis.

Multiple linear regression (MLR) and partial least squares regression (PLSR) were used to model the relation between the score data obtained from PARAFAC (X) and chemical data (Y). Full cross-validation was applied to all of the regression models. The regression models were evaluated using the adjusted $R^2$ and the root mean-square error of cross-validation (RMSECV) parameter.

The data analysis was carried out using Solo v. 5.0.1 (Eigenvector Research Inc., Manson, WA, USA) and Unscrambler 9.0 (CAMO, Oslo, Norway) software.

## 3. Results and Discussion

### 3.1. Evolution of Chemical Parameters of Cold-Pressed Rapeseed Oil During Storage

Freshly-pressed rapeseed oil samples were stored in the darkness and exposed to light in colorless and green glass bottles. As shown in Figure 1, colorless glass transmits radiation throughout the visible spectrum in the characteristic absorption range for both carotenoids and pheophytins. Green glass transmission is dependent on the spectral range and it is very low, below 500 nm, and higher in the long-wavelength range where long-wavelength absorption of pheophytins occurs.

The changes in the chemical parameters of the cold-pressed rapeseed oil stored for 6 months in the respective conditions are presented in Figure 2.

Table 1 reports the $p$-values of the analysis of covariance (ANCOVA) of the chemical data obtained for the oil samples stored for 6 months. The effect of time and storage conditions, as well as the results of the first-order interactions on chemical parameters, are presented. The results indicate that the storage time and conditions affected most of the studied oil quality parameters. The statistically significant effects of the storage conditions ($p < 0.05$) were observed for the time evolution of the following parameters: headspace oxygen, peroxide value (PV), $K_{270}$, polymerized triacylglycerols (TAGPs), diacylglycerols (DAG), tocopherols, and pheophytins.

The oxygen concentration in the bottle headspace and in the oil phase was measured in a non-invasive way using a fluorescence sensor. The headspace oxygen content was 20.9% in the freshly-bottled oil samples. A much lower concentration of 0.0007% was detected in the oil phase. The headspace oxygen content decreased during storage at rates depending on the storage conditions ($p = 0.014$), Figure 2a,b. The samples stored in colorless bottles exposed to light had the fastest drop in the headspace oxygen. The oil stored in green bottles revealed a similar decrease in the headspace oxygen concentration. The minimum headspace oxygen concentration, very close to 0.5%, was recorded in these samples when exposed to light in the third month of storage, with some increase observed at longer storage times. The smallest decrease in the headspace oxygen content was observed in oil stored in darkness; the lowest concentration of oxygen, at about 5%, was detected in these samples from the second to the fifth month of the experiment. Similar to the samples exposed to light, higher concentrations of the headspace oxygen were recorded at longer storage times. This minor increase in the oxygen concentrations in the final period of the experiment may result from the oxygen diffusion into the bottle or liberation of oxygen in the radical recombination reactions at the termination step of the oxidation process.

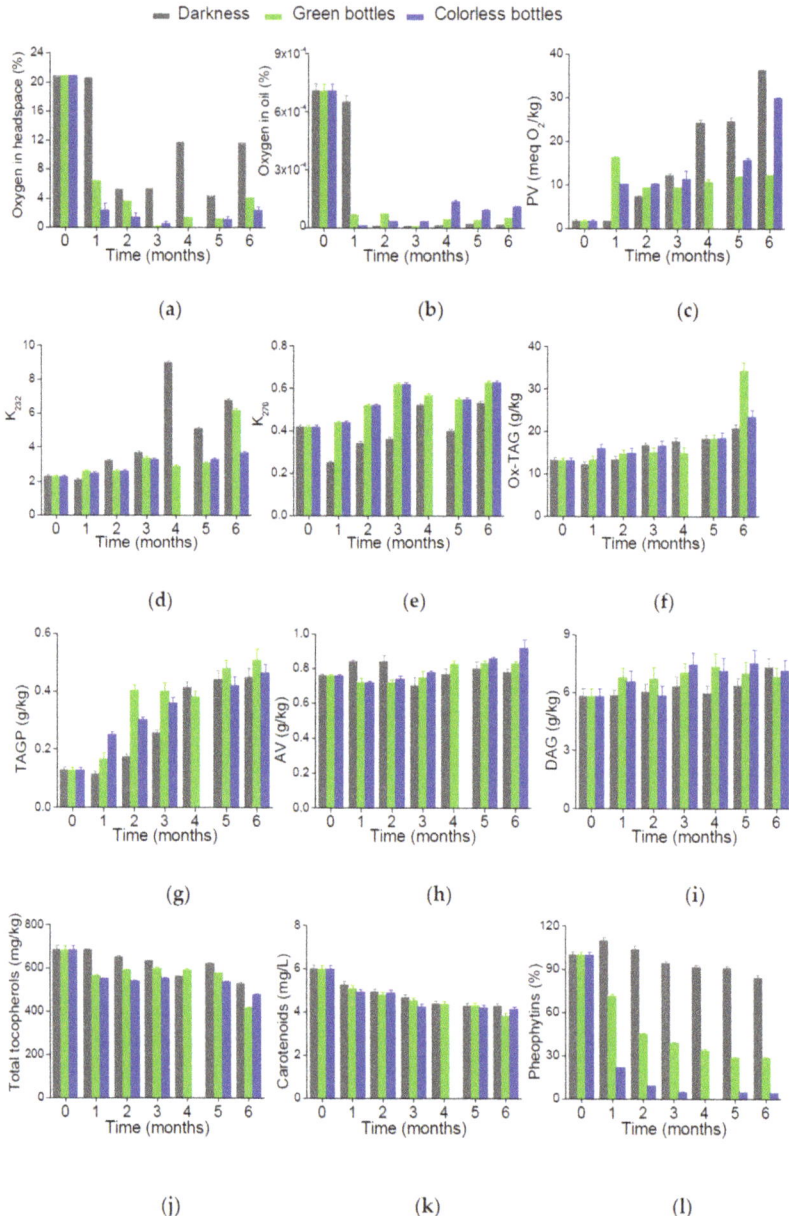

**Figure 2.** The changes of chemical parameters in the cold-pressed rapeseed oil during the 6 months of storage: (**a**) Oxygen concentration in the bottle headspace; (**b**) oxygen concentration in the oil; (**c**) peroxide value (PV); (**d**) specific absorption at 232 nm, $K_{232}$; (**e**) specific absorption at 270 nm, $K_{270}$; (**f**) content of the oxidized triacylglycerols (Ox-TAG); (**g**) content of polymerized triacylglycerols (TAGPs); (**h**) acid value (AV); (**i**) content of diacylglycerols (DAG); (**j**) total tocopherols; (**k**) carotenoids; (**l**) pheophytins.

**Table 1.** Results (*p*-values) of the analysis of covariance (ANCOVA) for the chemical parameters evaluated during the shelf-life of the rapeseed oil.

| Parameter | Time | Storage Conditions | Time*Storage Conditions |
| --- | --- | --- | --- |
| Oxygen in the bottle headspace | 0.378 | 0.014 | 0.707 |
| Oxygen in the oil | 0.210 | 0.616 | 0.081 |
| PV | <0.001 | 0.034 | <0.001 |
| $K_{232}$ | 0.009 | 0.089 | 0.292 |
| $K_{270}$ | 0.002 | 0.003 | 0.333 |
| Ox-TAG | 0.002 | 0.609 | 0.281 |
| TAGP | <0.001 | 0.039 | 0.181 |
| AV | 0.003 | 0.543 | 0.005 |
| DAG | 0.021 | 0.040 | 0.381 |
| Tocopherols | 0.006 | 0.022 | 0.577 |
| Carotenoids | <0.001 | 0.160 | 0.488 |
| Pheophytins | <0.001 | <0.001 | 0.105 |

Peroxide value (PV); acid value (AV); $K_{232}$, $K_{270}$—absorption coefficients at 232 and 270 nm, respectively; oxidised triacylglycerols (Ox-TAG), polymerized triacylglycerols (TAGPs), diacylglycerols (DAG).

Very low concentrations of oxygen dissolved in oil were recorded for the samples stored in darkness, starting from the second month of the experiment, as shown in Figure 2b.

The peroxide value (PV) evaluates the presence of the primary oxidation products of lipids (hydroperoxides). For a sample of fresh cold-pressed rapeseed oil, the PV was 1.8 meq $O_2$/kg and satisfied the requirements of the respective standards for cold-pressed oils [38]. The PV value changes during storage were significantly affected by the storage conditions ($p = 0.034$), as shown in Figure 2c. During the first month, the increase of the PV was much faster in the samples exposed to light (both in colorless and green bottles), as compared to the samples stored in darkness. The peroxide formation rates slowed down in the samples exposed to light after the first month. However, the PV increased almost linearly to the end of the experiment in the samples protected from light. The PV of 15 meq $O_2$/kg, allowed by the Codex Standard [38] for virgin oils, was exceeded in the samples stored in darkness after the third month.

The specific absorption coefficient at 232 nm ($K_{232}$) measures the concentration of conjugated dienes, another group of primary oxidation products. The $K_{232}$ value was lower in samples stored in darkness only after the first month ($p = 0.089$), as shown in Figure 2d. From the second month on, the concentration of the conjugated dienes was higher in the samples stored in darkness as compared to the respective samples exposed to light. However, there was no statistically significant effect of the storage conditions on the evolution of this parameter. The lipid hydroperoxides formed during autooxidation are conjugated dienes, whereas photooxidation leads to the formation of both conjugated and nonconjugated dienes [5].

The specific absorption coefficient at 270 nm ($K_{270}$), quantifying the conjugated trienes and the secondary oxidation products (carbonyl compounds), was significantly affected by the storage conditions ($p = 0.03$), as shown in Figure 2e. Its increase was more pronounced and similar in the samples stored in light, independently to the bottle glass color, as compared to those stored in darkness. These observations are in agreement with our previous study; we observed higher increases of $K_{270}$ for the olive oil samples stored under light as compared to those protected from light [39].

The observed changes in the PV, $K_{232}$, and $K_{270}$ parameters indicate that the oxidation was more advanced in the samples stored under light, thus the primary oxidation may have evolved into the secondary. These findings were further verified by the analysis of the evolution of the polar compounds. The substances that are typical oxidation (oxidized triacylglycerols (Ox-TAG), TAGP) and hydrolysis (DAG) products were quantified using the HPSEC method. The analysis of these compounds has already been successfully used for the estimation of the real extent of the oxidative and hydrolytic degradation of various edible oils and fats [40,41].

The evolution of the oxidation products (Ox-TAG and TAGP) is presented in Figure 2f,g. The oxidation product concentration was growing during the storage in all of the storage conditions. The formation of Ox-TAG, Figure 2f, ($p = 0.609$) was not significantly affected by the storage conditions. The Ox-TAG denomination comprises all of the oxidative products deriving from triacylglycerols, which can undergo further polymerization and degradation reactions. It was proposed that these substances could indicate the level of primary oxidation of oils [42].

TAGPs were initially formed at higher rates in the samples exposed to light ($p = 0.039$); however, after 4 months, similar levels of these compounds were observed in all samples, as shown in Figure 2g. The storage conditions significantly affected the evolution of these compounds. TAGPs were proposed as an index of the secondary oxidative degradation, because of their high stability and low volatility [42].

Hydrolytic degradation of oil during storage was evaluated on the basis of the acid value (AV) and DAG content. The AV increased in the samples stored under light throughout the entire duration of the experiment ($p = 0.543$), as shown in Figure 2h. Some fluctuations of the AV were recorded in darkness, although similar values were measured at the beginning and at the end of the storage period. The formation of DAG was affected by the storage conditions ($p = 0.040$), as shown in Figure 2i. Initially, it was higher in the samples exposed to light, while similar levels were recorded in all of the samples at the end of the experiment.

The progress of oil oxidation also depends on the presence of minor substances. Thus, the concentrations of tocopherols, carotenoids, and pheophytins were measured both in the fresh oil and during storage. Fresh oil was characterized by a relatively high content of vitamin E (total tocopherols) of 684 mg/kg. The tocopherol content decreased with the rates, depending on the storage conditions ($p = 0.022$), as shown in Figure 2j. The fastest decay was noted in the oil stored in colorless bottles. The degradation of tocopherols was the slowest in the samples stored in darkness. The lowest concentration of tocopherols after 6 months was recorded in the green glass bottle.

In fresh oil the content of carotenoid pigments was 6.0 mg/L. In contrast to tocopherols, the effect of the storage conditions on the decay of carotenoids during the experiment was insignificant ($p = 0.160$), as shown in Figure 2k.

The evolution of the pheophytins during the storage is of great importance due to their role of a photosensitizer in the photooxidation. In fresh oil, the content of pheophytin pigments of 1.7 mg/kg (pheophytin a) was rather low. The decay of pheophytins was markedly affected by the storage conditions ($p = 0.001$), as shown in Figure 2l. The effect of the bottle glass color was also clearly noticeable. The decay of pheophytins was the most advanced in the samples stored under light in the colorless bottles, as these pigments were partially protected from the photodegradation in the green glass bottles.

Based on the present results, we conclude that the oxidation of rapeseed oil is affected by light exposure mainly during the first few months of storage. Although the studied oil contained only minor amounts of pheophytins, even these speeded up its photooxidation. The green color of bottle glass slowed down the degradation of pheophytins and tocopherols at the initial period of the experiment, having a less pronounced effect on the oxidation during the entire period of storage. Moreover, it seems that oxygen headspace concentration was an important factor that limited the degree and rates of oxidation. It should be noted that due to the high rate of photooxidation, the study in the initial storage period, e.g., up to several days, should provide better insight into the kinetics of this process in colorless and green bottles.

### 3.2. Evolution of Fluorescence of Rapeseed Oil during Storage

The fluorescence methods provide two main advantages—high sensitivity and selectivity. Thus, fluorescence enables direct monitoring of minor components of oils, namely tocopherols, pheophytins, and oxidation products. Figure 3 presents the EEMs of fresh rapeseed oil and samples stored for 6 months, exposed to light in colorless and green bottles and stored in darkness. The oils with different oxidation degrees may by discriminated by the intensity of bands ascribed to the particular minor constituents.

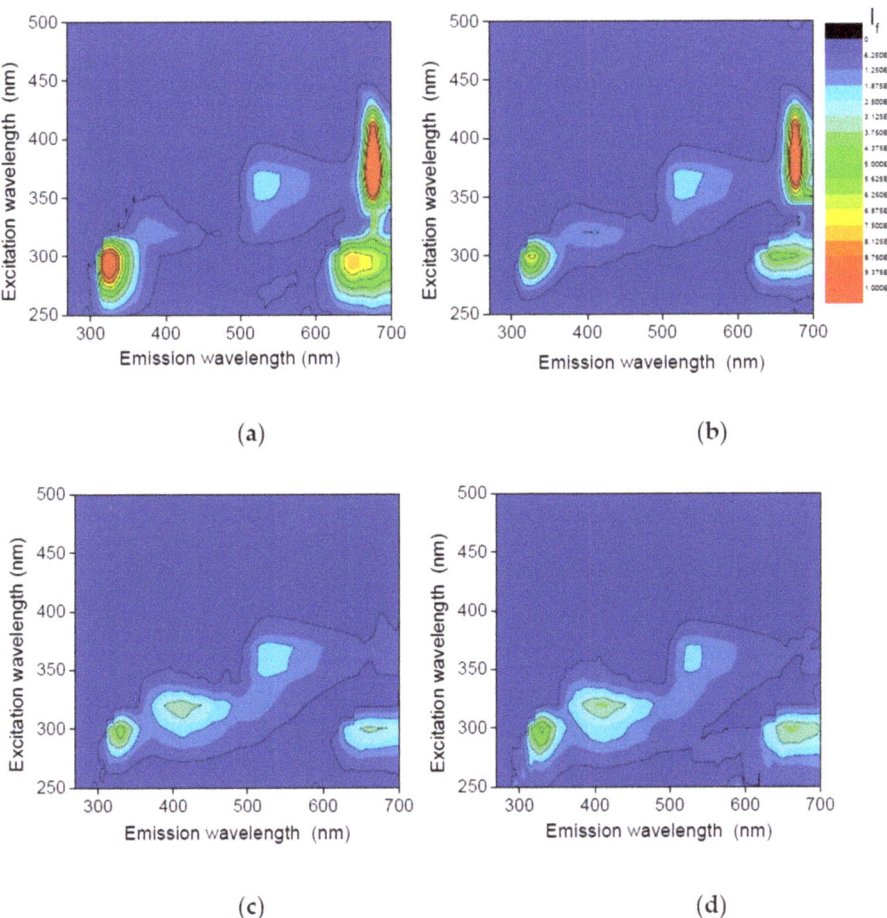

**Figure 3.** Total fluorescence spectra of freshly-pressed rapeseed oil, R0, (**a**) stored for 6 months in darkness, RD6 (**b**), and exposed to light in green, RG6, (**c**) and colorless, RL6, (**d**) bottles. The same intensity scale.

Marked differences in the shape and intensity of the fluorescence bands were observed between fresh and stored samples. The most intense fluorescent bands in the short- and long-wavelength regions in the fresh samples correspond respectively to tocopherols ($\lambda_{exc}/\lambda_{em}$ ca. 300/331 nm) and pheophytins (400/680 nm) [43]. The emission of phenolic compounds may also be observed at the short-wavelength side of this band. The intensity of these fluorescence bands decreased considerably in the stored samples.

The most pronounced differences between the fresh and stored oil spectra were observed in the intermediate spectral zone. Namely, a broad fluorescence band in the intermediate region ($\lambda_{exc}/\lambda_{em}$ ca. 320/400 nm) appeared during storage. The chemical compounds associated with the emission observed in oils in this range have not been unambiguously identified so far. However, in several studies it was suggested that this emission belongs to oxidation products [22,23,44]. Similarly, we attribute this emission to oxidation products, such as degradation products of pheophytins and/or polar compounds formed in the subsequent oxidation reactions.

The EEMs of all of the oil samples were investigated by means of the PARAFAC method with the objective to resolve the fluorescence landscapes into individual contributions of fluorescent compounds. Based on core consistency and a visual inspection of both the residuals and the loadings, an optimal PARAFAC model was estimated with four components (explained variance 99.1%, core consistency value 70).

Figure 4 presents the excitation and emission loadings of the four fluorescent components and the respective scores.

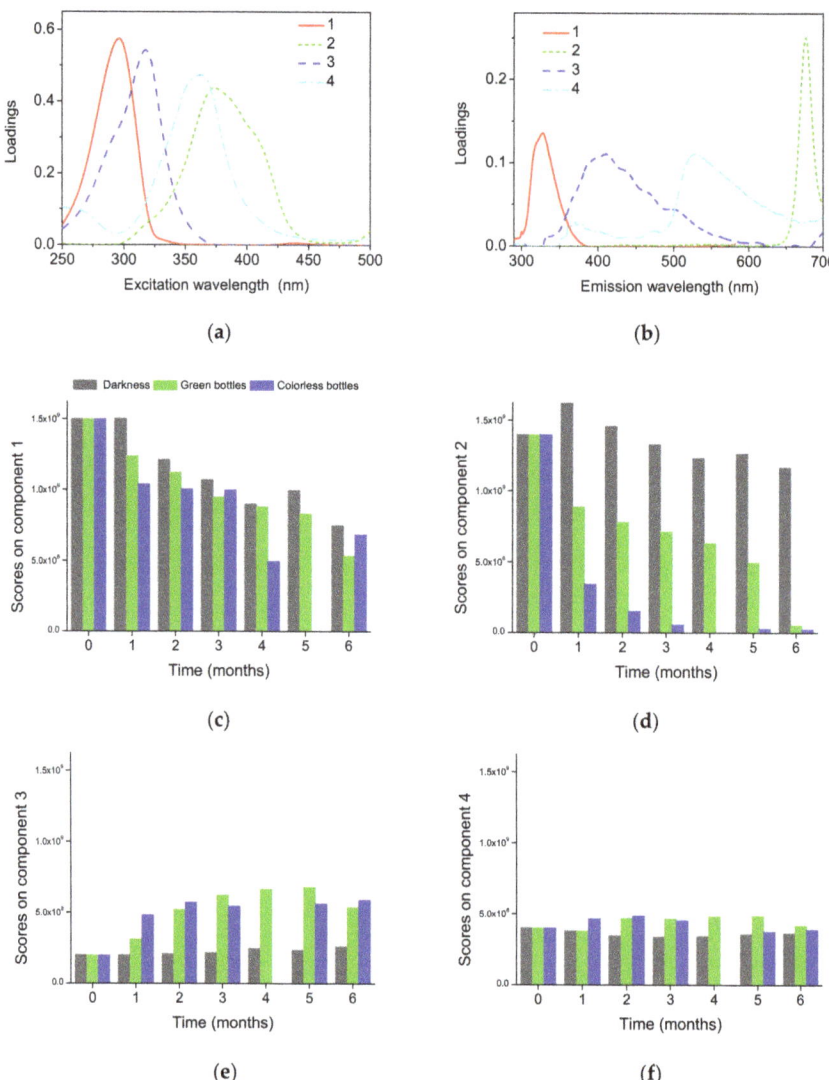

**Figure 4.** Results of parallel factor analysis (PARAFAC) of total fluorescence spectra of fresh and stored oil samples: Excitation (**a**) and emission (**b**) profiles. Scores vs storage time: scores on component 1 (**c**), scores on component 2 (**d**), scores on component 3 (**e**), scores on component 4 (**f**).

The first component appeared at 300/328 nm in excitation/emission and corresponds to tocopherols [43]. The second PARAFAC component appears at 370/677 nm, with a narrow emission band. This component may be ascribed to pheophytins [43]. The third PARAFAC component appears at 320/420 nm in excitation/emission, the fourth at 360/530 nm. The origin of component 3 may be ascribed to compounds formed during oil oxidation. The origin of component 4 was less obvious. The loading of an emission profile of this component, besides the main band, with the maximum at about 530 nm, exhibits additional broad emission with low intensity on the short-wavelength side. This may indicate that emission of some fluorescent components was not fully resolved, and therefore this component corresponds to the more than one chemical compounds.

We showed the contributions of each of the four PARAFAC components in Figure 4c–f. The score values obtained in the PARAFAC decomposition were plotted against the time to explore the progression of fluorescent components throughout the storage. The systematic variations of the score values, corresponding to the respective components, were observed. Thus, the decay of tocopherol (component 1) and pheophytin (component 2) in time was markedly affected by the storage conditions. The effect of bottle color was also noticeable. The decay of component 2 (pheophytin) was the most advanced in samples stored under light in colorless bottles. These pigments were partially protected from the photodegradation in the green bottles. The decay of tocopherol emission was influenced by light and bottle color, particularly in the first months of storage.

The appearance of the oxidation product fluorescence (component 3) was observed for samples stored in light after one month. The contribution of component 4 was affected by storage in a less systematic way. This component was presented in fresh oil and was slightly stronger in samples stored under light and weaker in samples stored in darkness, as compared to fresh oil.

### 3.3. Correlations between Chemical and Fluorescence Data

Next, we used principal component analysis (PCA) to visualize the relationship between differently stored oil samples and the variables describing their properties.

Figure 5 shows the results of the PCA for the analytical parameters and the contributions of the fluorescent components. The distribution of samples of rapeseed oil depended systematically on storage time and conditions, as shown in Figure 5a. Samples exposed to light and protected from light clearly follow different paths during the oxidation. The first two principal components (PC1 and PC2) described 74% of the total data variance.

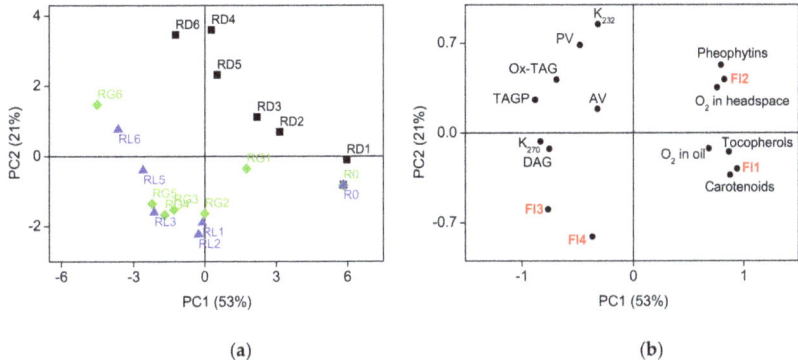

**Figure 5.** Results of principal component analysis (PCA) of chemical and fluorescent properties of oil samples: scores (**a**), correlation loadings (**b**). Fresh oil—R0; oils in colorless glass bottles, stored without light—RD; oils in colorless glass bottles exposed to light—RL; oils in green glass bottles, exposed to light—RG. Fl1, Fl2, Fl3, Fl4—fluorescent components obtained using PARAFAC. For the abbreviation of chemical parameters see Table 1.

The first principal component that explained 53% of the total data variance was linked to the time of storage. Samples spread along the PC1 axis, from positive to negative values, according to the storage time. The PC1 was positively correlated (correlation loadings ≥ 0.70) with the contents of tocopherols, carotenoids, and pheophytins, the oxygen concentration in the headspace and oil, and the first (tocopherols) and second (pheophytins) fluorescent components of the PARAFAC decomposition, as shown in Figure 5b. The PC1 was negatively correlated with the $K_{270}$, contents of polar compounds (TAGP, Ox-TAG, and DAG), and PARAFAC components 3. Thus, the progress of oxidation was accompanied by oxygen uptake and the decay of the minor components, tocopherols, carotenoids, and pheophytins, with simultaneous formation of the oxidation products.

The second principal component (explaining 21% of the total data variance) accounted for both storage time and the variability, due to the storage conditions. The PC2 was positively correlated with the PV and $K_{232}$ and negatively correlated with PARAFAC fluorescent component 4. The oxidative changes in samples of rapeseed oil during storage were clearly affected by storage conditions. Namely, the samples stored in darkness were characterized by higher concentrations of primary oxidation products, as evidenced by the PV, and $K_{232}$.

The significant correlations, evident from PCA analysis and calculated Pearson correlation coefficients ($r$), occurred between the analytical parameters and the contributions of the fluorescent components obtained in PARAFAC. The contribution of the first fluorescent component, identified as tocopherol, was correlated positively with tocopherols, determined by HPLC ($r = 0.83$), carotenoids ($r = 0.90$), pheophytins ($r = 0.63$), and oxygen concentration in oil ($r = 0.60$) and negatively with the formation of Ox-TAG ($r = -0.77$), TAGP ($r = -0.72$), DAG ($r = -0.63$), and the $K_{232}$ ($r = -0.63$) $K_{270}$ ($r = -0.77$). The decrease of the contribution of the second fluorescent component, identified as pheophytin, was correlated positively with the degradation of tocopherols ($r = 0.72$) and pheophytins ($r = 0.98$) and negatively with the increase in $K_{270}$ ($r = -0.68$) and DAG ($r = -0.62$). The contribution of the third fluorescent component, which increased during storage, was positively correlated with $K_{270}$ ($r = 0.72$) and DAG ($r = -0.64$) and negatively with pheophytins ($r = -0.91$) and oxygen in the headspace ($r = -0.64$)). In contrast, the fourth fluorescent component was only negatively correlated with pheophytins ($r = -0.68$).

The decays of the first and second fluorescent components were significantly correlated ($r = 0.69$). The degradation of the first component was correlated with the formation of the third component ($r = -0.58$), while the second component was negatively correlated with the third ($r = -0.85$) and fourth ($r = -0.59$) components. The positive correlation existed also between the third and fourth components ($r = 0.81$).

It should be noted that, among the discussed correlations between fluorescent components and analytical parameters, only those between component 1 and tocopherols and component 2 and pheophytins may be considered as direct. Other correlations were rather indirect and were a consequence of correlations between tocopherols and pheophytins content and respective analytical parameters.

In order to quantitatively model the relationship between the overall fluorescence characteristics (PARAFAC scores) and the chemical parameters, regression analysis was performed using MLR and PLSR methods. We tested the regression models for all of the studied parameters; however, we only present and discuss those models that had acceptable quality ($R^2_{cal} > 0.7$).

Table 2 presents the results of multivariate calibrations. The best calibration models were obtained for the prediction of pheophytin and total tocopherol content, using both MLR and PLSR on PARAFAC fluorescence scores.

**Table 2.** Results of multiple linear regression (MLR) and partial least squares regression (PLSR) of chemical parameters and fluorescent components extracted using PARAFAC.

| Parameter [1] | Regression Method | LV | $R^2_{cal}$ | $R^2_{cv}$ | RMSECV | RE (%) |
|---|---|---|---|---|---|---|
| $K_{232}$ | MLR |  | 0.774 | 0.625 | 1.1 | 29 |
|  | PLSR | 3 | 0.747 | 0.640 | 1.1 | 29 |
| Ox-TAG (g/kg) | MLR |  | 0.732 | 0.393 | 3.9 | 22 |
|  | PLSR | 2 | 0.678 | 0.498 | 3.7 | 21 |
| TAGP (g/kg) | MLR |  | 0.853 | 0.754 | 0.061 | 18 |
|  | PLSR | 3 | 0.853 | 0.799 | 0.059 | 17 |
| Tocopherols (mg/kg) | MLR |  | 0.903 | 0.825 | 27.2 | 4.7 |
|  | PLSR | 3 | 0.885 | 0.807 | 30.2 | 5.2 |
| Carotenoids (mg/L) | MLR |  | 0.861 | 0.735 | 0.25 | 5.4 |
|  | PLSR | 2 | 0.836 | 0.765 | 0.25 | 5.4 |
| Pheophytins (%) | MLR |  | 0.984 | 0.965 | 7.03 | 13.7 |
|  | PLSR | 3 | 0.984 | 0.972 | 6.65 | 12.4 |

[1] For the abbreviation of chemical parameters see Table 1. Latent variables (LVs) of PLSR models; $R^2$cal—determination coefficient of calibration; $R^2$cv—determination coefficient of validation; root mean-square error of cross-validation (RMSECV);— relative error (RE).

Satisfactory relations were also found between TAGP and fluorescence; however, the prediction error was quite high for this class of substances. The regression analysis did not give satisfactory results for Ox-TAG. A significant relationship was established for $K_{232}$, although these models had the highest relative errors of prediction.

Similarly to the observed correlations between individual fluorescent components and analytical parameters, quantitative relationships are also presented in Table 2, which may be considered as direct only for tocopherols and pheophytins. For other components, such as carotenoids, Ox-TAG, and TAGP, these relationships should be considered as indirect, and may be explained by the previously discussed correlations between various compounds involved in the oxidation processes.

The presented models confirmed that there is a quantitative relationship between fluorescence changes and some of chemical changes occurring during oxidation for the studied samples set. However, due to the indirect nature of these relationships, the respective models cannot be treated as universal models that enable valid determination of analytical parameters in other systems.

## 4. Conclusions

This study was aimed at investigating the potential of fluorescence in combination with chemometric methods for monitoring the rapeseed oil oxidation during storage in different conditions.

The results of chemical analyses revealed that the light exposure affected oxidation of rapeseed oil mainly in the initial period of storage. The minor amounts of pheophytins in oil studied speeded up its photooxidation. The green color of the glass bottle slowed down the degradation of pheophytins and tocopherols during the first few months and gave a less pronounced protective effect on the formation of oxidation products during the whole period of storage.

The PARAFAC of the front-face fluorescence excitation-emission matrices of rapeseed oil uniquely separated four fluorescent components, which had different evolution dynamics throughout the storage period, and revealed oxidative changes occurring in oil during storage. The fluorescence data were quantitatively related to some conventional chemical parameters describing the oxidation status of oil.

The present results show that fluorescence excitation-emission matrices associated with PARAFAC decomposition could be used for the direct monitoring of the oxidative degradation of rapeseed oils during storage.

**Author Contributions:** Conceptualization, E.S., K.W., W.K., A.G-Ś., I.K., T.G., F.C., V.M.P., C.S. and A.P.; data curation, E.S., K.W., A.G.-Ś., V.M.P., C.S. and A.P.; formal analysis, I.K. and T.G.; funding acquisition, E.S.; investigation, E.S., K.W., W.K., A.G.-Ś., V.M.P., C.S. and A.P.; methodology, E.S., K.W., W.K., A.G.-Ś., I.K., T.G., F.C.,

V.M.P., C.S. and A.P.; project administration, E.S.; supervision, E.S.; validation, E.S., K.W., W.K., A.G.-Ś., I.K., T.G., F.C., V.M.P., C.S. and A.P.; visualization, E.S.; writing—original draft, E.S.; writing—review and editing, E.S., K.W., W.K., A.G.-Ś., I.K., T.G., F.C., V.M.P., C.S. and A.P.

**Funding:** This research was funded by the Ministry of Science and Higher Education, Poland, grant number NN312428239. The APC was funded by Poznań University of Economics and Business.

**Conflicts of Interest:** The authors declare no conflict of interest.

## References

1. Matthäus, B.; Brühl, L. Why is it so difficult to produce high-quality virgin rapeseed oil for human consumption? *Eur. J. Lipid Sci. Technol.* **2008**, *110*, 611–617. [CrossRef]
2. Matthäus, B.; Brühl, L. Quality of cold-pressed edible rapeseed oil in Germany. *Nahrung* **2003**, *47*, 413–419. [CrossRef] [PubMed]
3. Tynek, M.; Pawłowicz, R.; Gromadzka, J.; Tylingo, R.; Wardencki, W.; Karlovits, G. Virgin rapeseed oils obtained from different rape varieties by cold pressed method—Their characteristics, properties, and differences. *Eur. J. Lipid Sci. Technol.* **2012**, *114*, 357–366. [CrossRef]
4. Yang, M.; Zheng, C.; Zhou, Q.; Huang, F.; Liu, C.; Wang, H. Minor components and oxidative stability of cold-pressed oil from rapeseed cultivars in China. *J. Food Compos. Anal.* **2013**, *29*, 1–9. [CrossRef]
5. Choe, E.; Min, D.B. Mechanisms and factors for edible oil oxidation. *Compr. Rev. Food Sci. F* **2006**, *5*, 169–186. [CrossRef]
6. Koski, A.; Pekkarinen, S.; Hopia, A.; Wähälä, K.; Heinonen, M. Processing of rapeseed oil: Effects on sinapic acid derivative content and oxidative stability. *Eur. Food Res. Technol.* **2003**, *217*, 110–114. [CrossRef]
7. Koski, A.; Psomiadou, E.; Tsimidou, M.; Hopia, A.; Kefalas, P.; Wähälä, K.; Heinonen, M. Oxidative stability and minor constituents of virgin olive oil and cold-pressed rapeseed oil. *Eur. Food Res. Technol.* **2002**, *214*, 294–298. [CrossRef]
8. Pawłowicz, R.; Gromadzka, J.; Tynek, M.; Tylingo, R.; Wardencki, W.; Karlovits, G. The influence of the UV irradiation on degradation of virgin rapeseed oils. *Eur. J. Lipid Sci. Technol.* **2013**, *115*, 648–658. [CrossRef]
9. Roszkowska, B.; Tańska, M.; Czaplicki, S.; Konopka, I. Variation in the composition and oxidative stability of commercial rapeseed oils during their shelf life. *Eur. J. Lipid Sci. Technol.* **2015**, *117*, 673–683. [CrossRef]
10. Rękas, A.; Ścibisz, I.; Siger, A.; Wroniak, M. The effect of microwave pretreatment of seeds on the stability and degradation kinetics of phenolic compounds in rapeseed oil during long-term storage. *Food Chem.* **2017**, *222*, 43–52. [CrossRef]
11. Rękas, A.; Siger, A.; Wroniak, M. The effect of microwave pre-treatment of rapeseed on the degradation kinetics of lipophilic bioactive compounds of the oil during storage. *Grasas y Aceites* **2018**, *69*, e233. [CrossRef]
12. Rękas, A.; Siger, A.; Wroniak, M.; Ścibisz, I. Phytochemicals and antioxidant activity degradation kinetics during long-term storage of rapeseed oil pressed from microwave-treated seeds. *Eur. J. Lipid Sci. Technol.* **2018**, *120*, 1700283. [CrossRef]
13. Wroniak, M.; Rękas, A. Nutritional value of cold-pressed rapeseed oil during long term storage as influenced by the type of packaging material, exposure to light & oxygen and storage temperature. *J. Food Sci. Technol.* **2016**, *53*, 1338–1347. [CrossRef] [PubMed]
14. Wu, Y.; Xu, F.; Ji, S.; Ji, J.; Qin, F.; Ju, X.; Wang, L. Influence of photooxidation on the lipid profile of rapeseed oil using UHPLC-QTOF-MS and multivariate data analysis. *Anal. Methods* **2019**, *11*, 2903–2917. [CrossRef]
15. Kanavouras, A.; Hernandez-Munoz, P.; Coutelieris, F.A. Packaging of olive oil: Quality issues and shelf life predictions. *Food Rev. Int.* **2006**, *22*, 381–404. [CrossRef]
16. Barriuso, B.; Astiasarán, I.; Ansorena, D. A review of analytical methods measuring lipid oxidation status in foods: A challenging task. *Eur. Food Res. Technol.* **2013**, *236*, 1–15. [CrossRef]
17. Kiritsakis, A.; Kanavouras, A.; Kiritsakis, K. Chemical analysis, quality control and packaging issues of olive oil. *Eur. J. Lipid Sci. Technol.* **2002**, *104*, 628–638. [CrossRef]
18. Cheikhousman, R.; Zude, M.; Bouveresse, D.J.-R.; Léger, C.L.; Rutledge, D.N.; Birlouez-Aragon, I. Fluorescence spectroscopy for monitoring deterioration of extra virgin olive oil during heating. *Anal. Bioanal. Chem.* **2005**, *382*, 1438–1443. [CrossRef]
19. Poulli, K.I.; Chantzos, N.V.; Mousdis, G.A.; Georgiou, C.A. Synchronous fluorescence spectroscopy: Tool for monitoring thermally stressed edible oils. *J. Agric. Food Chem.* **2009**, *57*, 8194–8201. [CrossRef]

20. Poulli, K.I.; Mousdis, G.A.; Georgiou, C.A. Monitoring olive oil oxidation under thermal and UV stress through synchronous fluorescence spectroscopy and classical assays. *Food Chem.* **2009**, *117*, 499–503. [CrossRef]
21. Sikorska, E.; Khmelinskii, I.V.; Sikorski, M.; Caponio, F.; Bilancia, M.T.; Pasqualone, A.; Gomes, T. Fluorescence spectroscopy in monitoring of extra virgin olive oil during storage. *Int. J. Food Sci. Technol.* **2008**, *43*, 52–61. [CrossRef]
22. Tena, N.; Garcia-Gonzalez, D.L.; Aparicio, R. Evaluation of virgin olive oil thermal deterioration by fluorescence spectroscopy. *J. Agric. Food Chem.* **2009**, *57*, 10505–10511. [CrossRef] [PubMed]
23. Tena, N.; Aparicio, R.; García-González, D.L. Chemical changes of thermoxidized virgin olive oil determined by excitation–emission fluorescence spectroscopy (EEFS). *Food Res. Int.* **2012**, *45*, 103–108. [CrossRef]
24. Guzmán, E.; Baeten, V.; Pierna, J.A.F.; García-Mesa, J.A. Evaluation of the overall quality of olive oil using fluorescence spectroscopy. *Food Chem.* **2015**, *173*, 927–934. [CrossRef]
25. Mbesse Kongbonga, Y.; Ghalila, H.; Majdi, Y.; Mbogning Feudjio, W.; Ben Lakhdar, Z. Investigation of heat-induced degradation of virgin olive oil using front face fluorescence spectroscopy and chemometric analysis. *J. Am. Oil Chem. Soc.* **2015**, *92*, 1399–1404. [CrossRef]
26. Saleem, M.; Ahmad, N.; Ali, H.; Bilal, M.; Khan, S.; Ullah, R.; Ahmed, M.; Mahmood, S. Investigating temperature effects on extra virgin olive oil using fluorescence spectroscopy. *Laser Phys.* **2017**, *27*, 125602. [CrossRef]
27. Domínguez Manzano, J.; Muñoz de la Peña, A.; Durán Merás, I. Front-face fluorescence combined with second-order multiway classification, based on polyphenol and chlorophyll compounds, for virgin olive oil monitoring under different photo- and thermal-oxidation procedures. *Food Anal. Methods* **2019**, *12*, 1399–1411. [CrossRef]
28. Hakonen, A.; Beves, J.E. Hue Parameter fluorescence identification of edible oils with a smartphone. *ACS Sens.* **2018**, *3*, 2061–2065. [CrossRef]
29. International Organization for Standardization. *ISO 3960:2007: Animal and Vegetable Fats and Oils—Determination of Peroxide Value—Iodometric (Visual) Endpoint Determination*; International Organization for Standardization: Geneva, Switzerland, 2007.
30. International Organization for Standardization. *ISO 660:2009: Animal and Vegetable Fats and Oils—Determination of Acid Value and Acidity*; International Organization for Standardization: Geneva, Switzerland, 2009.
31. International Organization for Standardization. *ISO 3656:2011: Animal and Vegetable Fats and Oils—Determination of Ultraviolet Absorbance Expressed as Specific UV Extinction*; International Organization for Standardization: Geneva, Switzerland, 2011.
32. Association of Official Analytical Chemists International. *Official Methods of Analysis of Association of Official Analytical Chemists International*; AOAC Press: Arlington, VA, USA, 2003.
33. Caponio, F.; Gomes, T.; Summo, C. Assessment of the oxidative and hydrolytic degradation of oils used as liquid medium of in-oil preserved vegetables. *J. Food Sci.* **2003**, *68*, 147–151. [CrossRef]
34. Gliszczyńska-Świgło, A.; Sikorska, E. Simple reversed-phase liquid chromatography method for determination of tocopherols in edible plant oils. *J. Chromatogr. A* **2004**, *1048*, 195–198. [CrossRef]
35. Zechmeister, L. *Cis-Trans Isomeric Carotenoids, Vitamins A, and Arylpolyenes*; Springer: Vienna, Austria, 1962.
36. American Oil Chemists' Society. *Official Methods and Recommended Practices of the American Oil Chemistry Society*; AOCS Press: Washington, DC, USA, 1993.
37. Andersen, C.M.; Bro, R. Practical aspects of PARAFAC modeling of fluorescence excitation-emission data. *J. Chemom.* **2003**, *17*, 200–215. [CrossRef]
38. Codex Alimentarius. *Codex Standard for Named Vegetable Oils*; Codex Stan. 210; Codex Alimentarius Comission: Rome, Italy, 1999.
39. Caponio, F.; Bilancia, M.; Pasqualone, A.; Sikorska, E.; Gomes, T. Influence of the exposure to light on extra virgin olive oil quality during storage. *Eur. Food Res. Technol.* **2005**, *221*, 92–98. [CrossRef]
40. Bilancia, M.T.; Caponio, F.; Sikorska, E.; Pasqualone, A.; Summo, C. Correlation of triacylglycerol oligopolymers and oxidised triacylglycerols to quality parameters in extra virgin olive oil during storage. *Food Res. Int.* **2007**, *40*, 855–861. [CrossRef]

41. Gomes, T.; Caponio, F.; Durante, V.; Summo, C.; Paradiso, V.M. The amounts of oxidized and oligopolymeric triacylglycerols in refined olive oil as a function of crude oil oxidative level. *LWT Food Sci. Technol.* **2012**, *45*, 186–190. [CrossRef]
42. Caponio, F.; Summo, C.; Bilancia, M.T.; Paradiso, V.M.; Sikorska, E.; Gomes, T. High performance size-exclusion chromatography analysis of polar compounds applied to refined, mild deodorized, extra virgin olive oils and their blends: An approach to their differentiation. *LWT Food Sci. Technol.* **2011**, *44*, 1726–1730. [CrossRef]
43. Zandomeneghi, M.; Carbonaro, L.; Caffarata, C. Fluorescence of vegetable oils: Olive oils. *J. Agric. Food Chem.* **2005**, *53*, 759–766. [CrossRef]
44. Guimet, F.; Ferré, J.; Boqué, R.; Rius, F.X. Application of unfold principal component analysis and parallel factor analysis to the exploratory analysis of olive oils by means of excitation–emission matrix fluorescence spectroscopy. *Anal. Chim. Acta* **2004**, *515*, 75–85. [CrossRef]

© 2019 by the authors. Licensee MDPI, Basel, Switzerland. This article is an open access article distributed under the terms and conditions of the Creative Commons Attribution (CC BY) license (http://creativecommons.org/licenses/by/4.0/).

Article

# The Honey Volatile Code: A Collective Study and Extended Version

Ioannis K. Karabagias *, Vassilios K. Karabagias and Anastasia V. Badeka

Laboratory of Food Chemistry, Department of Chemistry, University of Ioannina, 45110 Ioannina, Greece; vkarambagias@gmail.com (V.K.K.); abadeka@uoi.gr (A.V.B.)
* Correspondence: ikaraba@cc.uoi.gr or ioanniskarabagias@gmail.com; Tel.: +30-697-828-6866

Received: 23 September 2019; Accepted: 14 October 2019; Published: 17 October 2019

**Abstract:** Background: The present study comprises the second part of a new theory related to honey authentication based on the implementation of the honey code and the use of chemometrics. Methods: One hundred and fifty-one honey samples of seven different botanical origins (chestnut, citrus, clover, eucalyptus, fir, pine, and thyme) and from five different countries (Egypt, Greece, Morocco, Portugal, and Spain) were subjected to analysis of mass spectrometry (GC-MS) in combination with headspace solid-phase microextraction (HS-SPME). Results: Results showed that 94 volatile compounds were identified and then semi-quantified. The most dominant classes of compounds were acids, alcohols, aldehydes, esters, ethers, phenolic volatiles, terpenoids, norisoprenoids, and hydrocarbons. The application of classification and dimension reduction statistical techniques to semi-quantified data of volatiles showed that honey samples could be distinguished effectively according to both botanical origin and the honey code ($p < 0.05$), with the use of hexanoic acid ethyl ester, heptanoic acid ethyl ester, octanoic acid ethyl ester, nonanoic acid ethyl ester, decanoic acid ethyl ester, dodecanoic acid ethyl ester, tetradecanoic acid ethyl ester, hexadecanoic acid ethyl ester, octanal, nonanal, decanal, lilac aldehyde C (isomer III), lilac aldehyde D (isomer IV), benzeneacetaldehyde, *alpha*-isophorone, 4-ketoisophorone, 2-hydroxyisophorone, geranyl acetone, 6-methyl-5-hepten-2-one, 1-(2-furanyl)-ethanone, octanol, decanol, nonanoic acid, pentanoic acid, 5-methyl-2-phenyl-hexenal, benzeneacetonitrile, nonane, and 5-methyl-4-nonene. Conclusions: New amendments in honey authentication and data handling procedures based on hierarchical classification strategies (HCSs) are exhaustively documented in the present study, supporting and flourishing the state of the art.

**Keywords:** honey variety; honey code; HS-SPME/GC-MS; data handling; data bank; chemometrics

## 1. Introduction

The high consumer demand for authentic products along with the pressure on the market with products of low quality, distributed by cheap producing countries, creates the need for a multi-optional handling of natural-based products. A typical example of such products comprises honey—the sweet viscous solution obtained through the action of honeybees (*Apis mellifera*). The main types of honey include nectar and honeydew honeys. Nectar honeys are produced via the collection of the nectar of flowers by the honeybees.

On the other hand, honeydew honeys are characterized by the presence of secretions of plant-sucking insects (Hemiptera) living in the parts of the plants or conifer trees [1]. Given the historical meaning and symbolism of honey through the welfare of many civilizations [2], the latter has been subjected to exhaustive research. Apart from the basic components which are sugars and moisture, there are plenty of micro-constituents including minerals, phenolic compounds, organic acids, proteins, free amino acids, vitamins and volatile compounds and traces of lipid acids that have attracted researchers [3–7].

Among the aforementioned micro-constituents, volatile compounds are considered among the key parameters of honey sensory attributes. These contribute to the aroma providing the consumers with the emotional feelings related to regular consumption. It has been reported in the literature that volatile compounds of honey number in the hundreds, including esters, ethers, alcohols, acids, aldehydes, hydrocarbons, ketones, terpenes, norisoprenoids, carotenoid derivatives, furan and pyran derivatives, phenolic volatiles, benzene derivatives, quinones and other biomolecules originating from plants or bacteria metabolism with potential applications. The presence and quantity of these volatile compounds depends on the botanical and geographical origin of the honey [6–8].

The application of instrumental techniques has greatly favored the identification of the volatile compounds of honey. Numerous studies have been published using hydrodistillation, liquid–liquid extraction, simultaneous steam distillation extraction or Likens–Nickerson simultaneous distillation extraction micro-simultaneous steam distillation–solvent extraction, and ultrasonic solvent extraction for this purpose [7]. Some key volatile compounds that have been reported in the literature include benzene derivatives and phenolic volatiles for the case of strawberry tree honey [9]. Nonanal and cis-linalool oxide [2-[(2S,5R)-5-ethenyl-5-methyloxolan-2-yl]propan-2-ol] in combination with benzene derivatives and phenolic volatiles for Italian and Greek chestnut honeys [6,10]. Benzaldehyde, benzeneacetaldehyde and phenylethylalcohol were reported to be some characteristic volatile compounds of Spanish citrus and honeydew honeys [11,12]. The key volatile compounds of pine, fir, citrus, thyme, honeydew, and flower honeys harvested in Italy, Spain, Turkey, Greece, Morocco, and Brazil include benzaldehyde, benzeneacetaldehyde, octanal, nonanal, decanal, and different isomers of lilac aldehyde [3,4,6,8,13–16]. Norisoprenoids such as isophorone and 4-ketoisophorone (2,6,6-trimethyl-2-cyclohexene-1,4-dione) have been previously reported to serve as volatile markers of the floral origin of Sardinian strawberry tree and Indian saffron honeys [5,9]. *Alpha*-pinene, terpinolene, 2-phenylacetate and numerous other volatile compounds were considered as markers of the provenance of Argentinean honeys [17].

Based on the aforementioned, the objectives of the present study, which is collective in nature, were: (a) to classify clover, citrus, chestnut, eucalyptus, fir, pine, and thyme honeys from different countries (Egypt, Greece, Morocco, Spain, and Portugal) according to botanical origin based on the use of specific volatile compounds in combination with chemometrics and (b) classify honey samples according to the honey code—that is, a combination of the grammatical sequences of the different honey types used in the study by using the first letter of each honey type nomenclature. To the best of our knowledge, over the last 10 years, this is only the second study in the literature that implements, among others, a hierarchical classification strategy (HCS) for honey authentication [10], and the novelty of the study herein is highlighted by the use of a large number of different types of honey samples harvested in different parts of the world. Therefore, the whole procedure is more complicated and exhaustive "crash tests" are provided with the use of a multivariate analysis of variance (MANOVA), linear discriminant analysis (LDA), k-nearest neighbors (k-NN), and factor analysis (FA).

## 2. Materials and Methods

### 2.1. Honey Samples

One hundred and fifty-one honey samples ($n = 151$) were collected between the years 2011 and 2018 from Egypt, Greece, Morocco, Spain, and Portugal. Honey samples from Greece were obtained from Attiki Bee Culturing Co. Alex.Pittas S.A. (Athens, Greece); honey samples from Egypt, Spain and Morocco were obtained from local shops; honey samples from Portugal were obtained from APISMAIA (Povoa de Varzim, Portugal) The honey samples were subjected to volatile compound analysis according to botanical origin as clover (*Trifolium alexandrinum*), citrus (*Citrus* spp.), chestnut (*Castanea sativa*), eucalyptus (*Eucalyptus* spp.), fir (*Abies cephalonica*), Pine (*Pinus* spp.) and thyme (*Thymus capitatus* L. and *Thymus* spp.), which was confirmed by melissopalynological analysis [16]. In particular, clover honeys ($n = 8$) originated from Egypt; citrus honeys originated from Egypt ($n = 7$),

Spain ($n = 8$), Morocco ($n = 6$), and Greece ($n = 10$); chestnut honeys originated from Greece ($n = 1$) and Portugal ($n = 3$); eucalyptus honeys ($n = 4$) originated from Portugal; fir honeys ($n = 31$) originated from Greece; pine honeys ($n = 39$) originated from Greece; thyme honeys ($n = 42$) originated from Egypt ($n = 7$), Greece ($n = 12$), Morocco ($n = 6$), and Spain ($n = 10$). Honey samples were shipped to the laboratory and maintained firstly at room temperature in paper boxes for sampling which was started at once. Sampling and analysis followed the sequence of honey type harvesting through the aforementioned years. The paper boxes were kept away from UV light. The quantity of honey samples left was stored at $4 \pm 1\,°C$.

## 2.2. Honey Code Development

The honey code was used to construct the group of objects that would be subjected to statistical analysis using the first letter of each honey type. Clover, citrus and chestnut honeys ($n = 42$) were represented by CCC; eucalyptus honeys ($n = 4$) by E; fir honeys ($n = 31$) by F; pine honeys ($n = 39$) by P, and thyme honeys ($n = 42$) by T. The main purpose of this hierarchical procedure was to test the classification ability of honey samples from different countries based on the use of specific volatile compounds and chemometrics according to honey type lettering (use of the first letter) [10].

## 2.3. Analysis of Gas Chromatography–Mass Spectrometry in Combination with Headspace Solid-Phase Microextraction (HS-SPME/GC–MS)

The experimental strategy for the isolation and semi-quantification of honey volatile compounds along with HS-SPME/GC–MS equipment and analysis conditions are given in details in previous work [10]. The mass spectral library used for the identification of volatile compounds was Wiley 7 (2005) of the National Institute of Standards and Technology (NIST). Only the volatile compounds that had ≥80% similarity with those of the Wiley MS library were tentatively identified using the GC–MS spectra. Data were expressed as concentration (Canalyte, mg/kg) based on the ratio of peak areas of the isolated volatile compounds to that of the internal standard (benzophenone) multiplied by the final concentration of the internal standard, assuming a response factor (RF) equal to 1 for all the compounds. An additional method of identification was considered and included the calculation of Kovats indices using a mixture of $n$-alkanes (C8–C20) which was supplied by Supelco (Bellefonte, PA, USA). The standard mixture was dissolved in $n$-hexane. The retention time of the standards was determined according to the temperature-programmed run used in the analysis of honey samples. MS and Kovats indices data were compared to those found in the Wiley MS library. A solvent delay of 5 min was inserted in the program, in order to avoid the elution of ethanol, in which the internal standard was dissolved. Each sample was analyzed in duplicate and the results were averaged.

## 2.4. Statistical Analysis

Multivariate analysis of variance (MANOVA), linear discriminant analysis (LDA), k-nearest neighbors (k-NN), and factor analysis (FA) were applied to the semi-quantitative data of volatile compounds. The first part of the statistical analysis included the botanical origin differentiation of clover, citrus, chestnut, eucalyptus, fir, pine, and thyme honeys based on the use of the significant volatiles ($p < 0.05$) which served as the independent variables. The botanical origin served as the grouping variable consisted of 7 groups (clover, citrus, chestnut, eucalyptus, fir, pine, and thyme honeys). In the second part of the statistical analysis, honey samples were grouped according to the honey code. The expectation was to investigate whether classification results could be improved.

For the MANOVA analysis, the indices of the multivariate hypothesis such as Pillai's Trace, Wilks' Lambda, Hotelling's Trace, and Roy's Largest Root were computed to determine whether there was a multivariate effect of volatile compounds ($p < 0.05$) on the botanical origin or the honey code of honey. The size of the effect was further evaluated by consideration of partial eta squared values ($\eta^2$). It should also be stressed that the lower the value of Wilks' Lambda, the higher the differences between groups of objects.

Considering only the significant volatiles, LDA was then applied to classify honey samples according to group membership based on the use of original and cross-validation methods. LDA provides linear discriminant functions originating from the combinations of the significant variables (all independent variables are entered together/simultaneously in the analysis) multiplied by the standardized canonical discriminant function coefficients plus a constant, characteristic for each discriminant function [18]. Moreover, the tolerance test was also computed in the analysis. Tolerance may be defined as the proportion of a variable's variance not accounted for by other independent variables in the created discriminant function. It practically shows that a variable with very low tolerance contributes little information to the predictive model and may cause computational problems [19].

For the k-nearest neighbors analysis, the botanical origin of samples or the honey code served as the target parameter, while the significant volatiles ($p < 0.05$) served as the features. The number of the k-nearest neighbors was set by default equal to the higher number provided by the SPSS software—that is, 3–5. The classification ability of the constructed model was estimated by the application of training and holdout partitions. In the training sample, 70% of the cases were randomly assigned to partitions, while the rest of the cases were assigned to the holdout sample. The distances of an unknown object from all the members of the training set were calculated using the Euclidean distance in multi-dimensional space. The k-smallest distances between the unknown object and the training set sample were then identified. K is normally a small odd number, and the unknown variable is allocated to the class with the majority of these k distances [10]. For the performance of feature selection among the population of significant variables, k-NN analysis was run again using only the specified predictors that built the model (usually 3–5) in the previous step, in order to reduce the number of predictors to those having the lower error rate and k distance.

Afterwards, a data mining technique such as factor analysis (FA) was applied to the significant parameters in order to explain the total variance of the constructed model in the multidimensional space. At the same time, FA provides a reduction in the variables, in order to gain the most pronounced ones with the higher correlation and communalities of independent latent variables. The communalities indicate the common variance shared by factors with specific variables. A higher communality indicates that a larger amount of the variance in the variable has been extracted by the factor solution. For the most effective data collection during FA, communalities should be ≥0.4. The extraction method was principal component analysis (PCA). The rotation method used was Varimax with Keiser Normalization. The Varimax rotation is used in statistical analysis to simplify the expression of a particular sub-space in terms of just a few major items each. The actual coordinate system (practically constant) is the orthogonal basis that is being rotated to align with these coordinates. The sub-space can be defined with either PCA or FA. Varimax maximizes the sum of the variances of the squared loadings (squared correlations between variables and factors [20].

The accuracy and strength of factor analysis was supported further by the Kaiser–Meyer–Olkin (KMO) test, which comprises a measure of how well suited the data is for factor analysis. The acceptable value considered was that of KMO ≥ 0.50. In addition, the effectiveness and suitability of factor analysis was explored using Bartlett's test of sphericity. This test highlights the hypothesis that the correlation matrix is an identity matrix, which would indicate that the variables incorporated into the model are unrelated and therefore unsuitable for structure detection. Small probability values ($p < 0.05$) indicate that a factor analysis may be useful with data treatment [20]. Statistical analysis was run using the SPSS (version 20.0, IBM, Armonk, NY, USA) statistics software.

## 3. Results and Discussion

### 3.1. Volatile Compounds of Clover, Citrus, Chestnut, Eucalyptus, Fir, Pine, and Thyme Honeys

A considerable number of volatile compounds were putatively identified and semi-quantified. In total, 94 volatile compounds of different classes were isolated. Table 1 lists the volatile compounds according to retention time and their class. The majority of volatile compounds (62 volatiles) varied significantly ($p < 0.05$) according to the botanical origin of honey.

Figure 1 shows a typical chromatogram of clover honey from Egypt, indicating with numbers some selected key volatile compounds. In supplementary material (Figures S1–S6), typical chromatograms of citrus, chestnut, eucalyptus, fir, pine, and thyme honeys from the investigated regions are given.

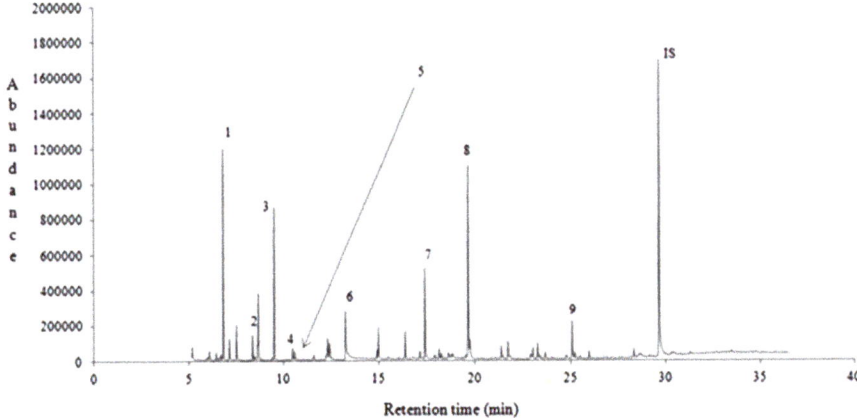

**Figure 1.** A typical gas chromatogram of clover honey (no. 3) from Egypt indicating selected key volatile compounds. 1: 2-methyl-Butanal. 2: 3-methyl-Butanal. 3: Heptane. 4: 3-methyl-1-Butanol. 5: 2-methyl-1-Butanol. 6: Furfural. 7: Octanal. 8: Nonanal. 9: Decanoic acid ethyl ester. IS: Internal standard.

Clover honeys showed higher amounts (mg/kg) of 2-methylbutanal, 3-methylbutanal, 3-methyl-1-butanol, and 2-methyl-1-butanol compared to the other honey types. The aforementioned compounds comprise isoleucine- and leucine-derived volatiles and are found in a wide range of foods such as honey, beer, cheese, coffee, chicken, fish, chocolate, olive oil, and tea. These volatile compounds are associated with a fruity and malty flavor [21,22].

Citrus honeys were characterized by the higher amounts (mg/kg) of lilac aldehyde C (isomer III), dill ether, methylanthranilate and herboxide (isomer II). These compounds have been reported previously to dominate the volatile profile of citrus honeys among other honey types [3,4].

Chestnut honeys showed the highest amounts (mg/kg) of benzaldehyde, benzeneacetaldehyde, 1-octene, and furfural. The latter volatile compound may serve as an indicator of heat resistance/treatment of chestnut honeys.

Eucalyptus honeys were characterized by the presence of heptane and *beta*-damascenone, since these compounds were found in higher amounts (mg/kg). The presence of hydrocarbons in honey is a typical phenomenon, whereas *beta*-damascenone is a cyclic carotenoid derivative and possesses a rose odor [22].

Fir honeys had higher amounts (mg/kg) of nonanal, decanal, hexanoic acid ethyl ester, heptanoic acid ethyl ester, octanoic acid ethyl ester, nonanoic acid ethyl ester, decanoic acid ethyl ester, dodecanoic acid ethyl ester, tetradecanoic acid ethyl ester, nonane, 5-methyl-4-nonene, 6-methyl-5-hepten-2-one, geranyl acetone, 1-2-(furanyl)-ethanone, *alpha*-isophorone, 4-ketoisophorone, and 2-hydroxyisophorone. The volatile compounds 6-methyl-5-hepten-2-one and geranyl acetone comprise open-chain carotenoid-derived volatiles, whereas the isophorone related compounds belong to the class of norisoprenoids, resulting from the degradation of terpenoids [22]. Geranyl acetone and 6-methyl-5-hepten-2-one possess a strong floral, green and fruit-like odor. In a previous study, these compounds were characterized as exocrine products (cephalic secretions) of cleptoparasitic bees (Holcopasites) [23].

Table 1. Semi-quantitative results of volatile compounds tentatively identified in honey samples according to botanical origin.

| RT [a] (min) | Volatile Compounds (mg/kg) | RIexp | Clover | Citrus | Chestnut | Eucalyptus | Fir | Pine | Thyme | F | p |
|---|---|---|---|---|---|---|---|---|---|---|---|
| **Acids** | | | | | | | | | | | |
| 5.40 | Formic acid | <800 | ni | 0.02 (0.09) | ni | ni | ni | 0.01 (0.04) | 0.02 (0.09) | 0.312 | 0.930 |
| 6.53 | Acetic acid | <800 | 0.09 (0.14) | 0.04 (0.16) | ni | ni | 0.07 (0.22) | 0.16 (0.65) | 0.05 (0.17) | 0.570 | 0.753 |
| 13.02 | Pentanoic acid | 823 | 0.06 (0.05) | ni | ni | ni | ni | 0.0004 | 0.002 | 3.034 | 0.008 |
| 16.36 | Hexanoic acid | 956 | 0.12 (0.11) | ni | ni | ni | ni | 0.01 (0.06) | ni | 0.572 | 0.752 |
| 18.57 | Heptanoic acid | 1053 | ni | ni | ni | ni | ni | 0.0001 (0.0005) | ni | 0.468 | 0.831 |
| 20.65 | Octanoic acid | 1151 | 0.05 (0.09) | ni | ni | ni | ni | 0.03 (0.13) | ni | 1.097 | 0.367 |
| 22.60 | Nonanoic acid | 1249 | ni | ni | ni | ni | ni | 0.03 (0.11) | ni | 1.722 | 0.120 |
| 24.42 | Decanoic acid | 1348 | 0.09 (0.16) | ni | ni | ni | ni | 0.01 (0.03) | ni | 1.624 | 0.144 |
| 27.79 | Dodecanoic acid | 1549 | ni | ni | ni | ni | ni | 0.002 (0.004) | ni | 3.488 | 0.003 |
| **Aldehydes** | | | | | | | | | | | |
| 6.80 | 2-methyl-1-Butanal | <800 | ni | 0.02 (0.05) | ni | ni | ni | ni | 0.04 (0.17) | 1.876 | 0.089 |
| 8.35 | 3-methyl-1-Butanal | <800 | ni | 0.02 (0.01) | ni | ni | ni | ni | 0.01 (0.05) | 6.548 | 0.000 |
| 13.36 | Furfural | 836 | ni | 0.06 (0.14) | 0.95 (1.57) | ni | 0.62 (0.31) | 0.12 (0.51) | 0.01 (0.02) | 12.404 | 0.000 |
| 16.63 | 5-methyl-2-Furaldehyde | 967 | ni | 0.002 (0.011) | 0.27 (0.23) | ni | 0.22 (0.31) | 0.013 (0.057) | 0.005 (0.02) | 0.685 | 0.662 |
| 16.86 | Benzaldehyde | 977 | ni | 0.02 (0.03) | ni | 0.07 (0.08) | ni | 0.05 (0.18) | 0.07 (0.07) | 4.966 | 0.000 |
| 17.51 | Octanal | 1005 | ni | 0.003 (0.01) | 0.26 (0.34) | 0.03 (0.03) | 0.36 (0.34) | 0.08 (0.40) | 0.01 (0.02) | 6.607 | 0.000 |
| 18.66 | Benzeneacetaldehyde | 1059 | 0.05 (0.09) | 0.06 (0.10) | 0.38 (0.33) | 0.05 (0.05) | 0.31 (0.40) | 0.05 (0.14) | 0.31 (0.44) | 5.009 | 0.000 |
| 19.75 | Nonanal | 1107 | 0.09 (0.16) | 0.08 (0.09) | 1.29 (2.01) | ni | 3.30 (3.28) | 0.17 (0.35) | 0.03 (0.05) | 18.021 | 0.000 |
| 20.58 | Lilac aldehyde A (isomer I) | 1145 | ni | 0.09 (0.16) | ni | ni | ni | ni | 0.01 (0.03) | 6.584 | 0.000 |
| 20.60 | Lilac aldehyde B (isomer II) | 1154 | ni | 0.11 (0.17) | ni | 0.01 (0.02) | ni | ni | 0.003 (0.01) | 8.820 | 0.000 |
| 20.78 | Lilac aldehyde C (isomer III) | 1172 | ni | 0.34 (0.32) | 0.20 (0.34) | ni | ni | ni | 0.02 (0.03) | 19.726 | 0.000 |
| 21.23 | Lilac aldehyde D (isomer IV) | 1178 | ni | 0.01 (0.03) | 0.10 (0.17) | ni | ni | ni | ni | 9.517 | 0.000 |
| 21.83 | Decanal | 1209 | ni | 0.02 (0.05) | 0.89 (1.44) | 0.03 (0.02) | 1.43 (1.28) | ni | 0.01 (0.02) | 22.745 | 0.000 |
| 22.86 | 2-methyl-3-phenylPropanal | 1245 | ni | ni | ni | ni | ni | ni | 0.004 (0.03) | 0.542 | 0.775 |
| 23.06 | 4-methoxy-Benzaldehyde | 1252 | ni | ni | ni | ni | ni | ni | 0.02 (0.01) | 0.542 | 0.775 |
| 27.11 | 5-methyl-2-phenyl-2-Hexenal | 1475 | ni | ni | ni | ni | ni | ni | 0.004 (0.001) | 2.936 | 0.010 |
| **Alcohols** | | | | | | | | | | | |
| 10.45 | 3-methyl-1-Butanol | <800 | 0.034 (0.05) | 0.006 (0.034) | ni | ni | ni | ni | 0.01 (0.06) | 1.436 | 0.205 |
| 10.56 | 2-methyl-1-Butanol | <800 | 0.02 (0.04) | ni | ni | ni | ni | ni | 0.001 (0.003) | 6.671 | 0.000 |
| 16.80 | 1-Octen-3-ol | 979 | ni | ni | ni | ni | ni | ni | 0.42 (1.17) | 2.396 | 0.031 |
| 17.20 | 3-Octanol | 992 | ni | ni | ni | ni | ni | ni | 0.01 (0.06) | 0.542 | 0.775 |
| 17.91 | 2-ethyl-1-Hexanol | 1027 | ni | 0.001 (0.005) | 0.04 (0.04) | 0.13 (0.06) | 0.30 (0.32) | 0.05 (0.16) | 0.001 (0.002) | 11.588 | 0.000 |
| 18.94 | 1-Octanol | 1069 | ni | ni | 0.10 (0.18) | ni | 0.15 (0.22) | 0.02 (0.08) | ni | 7.271 | 0.000 |
| 20.15 | Phenylethylalcohol | 1129 | ni | ni | ni | ni | ni | ni | 0.05 (0.10) | 4.820 | 0.000 |
| 21.07 | 1-Nonanol | 1171 | ni | ni | 0.47 (0.55) | 0.04 (0.05) | ni | 0.14 (0.45) | ni | 3.271 | 0.005 |
| 23.02 | 1-Decanol | 1271 | ni | ni | ni | ni | 0.08 (0.24) | 0.001 (0.004) | ni | 2.258 | 0.041 |
| **Esters** | | | | | | | | | | | |
| 5.19 | Formic acid ethyl ester | <800 | ni | ni | ni | ni | ni | ni | 0.006 (0.001) | 3.559 | 0.003 |
| 7.18 | Acetic acid ethyl ester | <800 | 0.13 (0.18) | 0.01 (0.02) | ni | ni | ni | ni | ni | 11.814 | 0.000 |
| 17.16 | Hexanoic acid ethyl ester | 996 | ni | ni | 0.09 (0.10) | 0.08 (0.03) | 0.28 (0.34) | 0.01 (0.01) | 0.02 (0.04) | 11.742 | 0.000 |
| 19.47 | Heptanoic acid ethyl ester | 1097 | ni | ni | 0.04 (0.03) | 0.01 (0.02) | 0.21 (0.31) | 0.07 (0.01) | ni | 8.949 | 0.000 |
| 21.38 | Octanoic acid ethyl ester | 1193 | ni | 0.01 (0.02) | 0.46 (0.37) | 0.22 (0.16) | 1.08 (0.42) | 0.10 (0.21) | 0.07 (0.10) | 80.095 | 0.000 |
| 23.27 | Nonanoic acid ethyl ester | 1298 | ni | 0.03 (0.04) | 0.69 (0.71) | 0.28 (0.10) | 2.11 (1.14) | 0.19 (0.44) | 0.05 (0.07) | 49.997 | 0.000 |

[a] RT: Retention time. RIexp: Experimental retention index values using standard hydrocarbons naturally present in honey. ni: not identified. The non-identified volatile compounds were treated as zeros for chemometrics and not as missing values. F: Values of the F distribution. p: probability.

**Table 1.** Cont.

| RT [a] (min) | Volatile Compounds (mg/kg) | RIexp | Clover | Citrus | Chestnut | Eucalyptus | Fir | Pine | Thyme | F | p |
|---|---|---|---|---|---|---|---|---|---|---|---|
| 24.74 | Methyl anthranilate | 1366 | ni | 0.02 (0.04) | ni | ni | ni | ni | ni | 5.918 | 0.000 |
| 25.05 | Decanoic acid ethyl ester | 1389 | 0.04 (0.06) | 0.02 (0.05) | 0.48 (0.63) | 0.05 (0.06) | 0.99 (0.55) | 0.07 (0.15) | 0.03 (0.06) | 49.415 | 0.000 |
| 26.87 | Undecanoic acid ethyl ester | 1491 | ni | ni | 0.02 (0.03) | ni | 0.02 (0.03) | ni | ni | 7.569 | 0.000 |
| 28.46 | Dodecanoic acid ethyl ester | 1591 | ni | ni | 0.22 (0.28) | 0.05 (0.01) | 0.59 (0.38) | 0.03 (0.07) | ni | 29.067 | 0.000 |
| 29.96 | Tridecanoic acid ethyl ester | 1791 | ni | ni | 0.23 (0.40) | ni | 0.04 (0.19) | 0.02 (0.10) | ni | 2.358 | 0.033 |
| 31.31 | Tetradecanoic acid ethyl ester | 1883 | ni | ni | 0.14 (0.24) | 0.004 (0.01) | 0.22 (0.40) | ni | ni | 5.991 | 0.000 |
| 34.78 | Hexadecanoic acid ethyl ester | 1982 | ni | ni | 0.02 (0.02) | 0.01 (0.01) | ni | ni | ni | 35.197 | 0.000 |
| **Ethers** | | | | | | | | | | | |
| 21.72 | Dill ether | 1203 | ni | 0.07 (0.07) | ni | ni | ni | ni | 0.01 (0.02) | 18.039 | 0.000 |
| 22.27 | Thymol methyl ether [Benzene, 2-methoxy-4-methyl-1-(1-methylethyl)-] | 1235 | ni | ni | ni | ni | ni | ni | 0.10 (0.27) | 2.534 | 0.023 |
| 22.48 | Carvacrol methyl ether | 1246 | ni | ni | ni | ni | ni | ni | 0.05 (0.17) | 1.797 | 0.104 |
| 29.52 | Octyl ether | 1617 | ni | ni | ni | ni | ni | ni | 0.02 (0.08) | 1.374 | 0.229 |
| **Hydrocarbons** | | | | | | | | | | | |
| 9.49 | Heptane | <800 | ni | 0.04 (0.05) | 0.26 (0.22) | 0.41 (0.11) | ni | ni | 0.11 (0.18) | 21.131 | 0.000 |
| 12.18 | 1-Octene | <800 | ni | ni | 0.36 (0.31) | ni | 0.30 (0.36) | 0.15 (0.89) | ni | 1.616 | 0.147 |
| 12.27 | Octane | 800 | ni | 0.01 (0.02) | 0.66 (0.32) | 0.96 (0.59) | ni | 1.35 (8.01) | 0.04 (0.08) | 0.512 | 0.798 |
| 14.81 | 1-Nonene | 889 | ni | ni | ni | ni | ni | 0.18 (0.98) | ni | 0.615 | 0.718 |
| 15.02 | Nonane | 900 | ni | ni | 0.20 (0.27) | 0.05 (0.04) | 0.53 (0.36) | 0.18 (0.98) | ni | 3.681 | 0.002 |
| 16.98 | 5-methyl-4-Nonene | 981 | ni | ni | 0.06 (0.11) | ni | 0.25 (0.29) | 0.001 (0.003) | ni (0.001) | 13.789 | 0.000 |
| 17.39 | Decane | 1000 | ni | ni | 0.03 (0.02) | 0.03 (0.01) | ni | 0.002 (0.003) | 0.001 (0.003) | 76.961 | 0.000 |
| 19.61 | Undecane | 1100 | ni | ni | ni | ni | ni | ni | ni | 2.132 | 0.053 |
| **Ketones** | | | | | | | | | | | |
| 15.37 | 1-(2-furanyl)-Ethanone | 914 | ni | ni | 0.08 (0.14) | ni | 0.14 (0.32) | ni | 0.01 (0.05) | 3.546 | 0.003 |
| 16.76 | 6-methyl-5-Hepten-2-one | 973 | ni | ni | 0.01 (0.01) | 0.01 (0.01) | 0.02 (0.04) | ni | ni | 4.139 | 0.001 |
| 17.97 | (1S,4S,5R)-4-methyl-1-propan-2-ylbicyclo(3.1.0)-hexan-3-one (beta-Thujone) | 1125 | ni | ni | ni | ni | ni | ni | ni | 2.134 | 0.053 |
| 21.10 | (1R,4S)-1,7,7-trimethylbycyclo-(2.2.1)-heptan-2-one (Camphor) | 1143 | ni | ni | ni | ni | ni | 0.04 (0.01) | 0.08 (0.21) | 2.425 | 0.029 |
| 24.83 | 3-hydroxy-4-phenyl-Butanone | 1347 | ni | ni | ni | ni | ni | ni | 0.01 (0.03) | 1.851 | 0.093 |
| 25.37 | 1-(2,6,6-trimethyl-1,3-cyclohexadien-1-yl)-2-Buten-1-one (beta-Damascenone) | 1359 | ni | ni | 0.01 (0.01) | 0.08 (0.04) | 0.01 (0.03) | 0.03 (0.14) | 0.002 (0.003) | 1.102 | 0.364 |
| 26.27 | (E)-6,10-dimethyl-5,9-Undecadien-2-one (Geranyl acetone) | 1455 | ni | ni | ni | ni | 0.01 (0.03) | ni | ni | 5.276 | 0.000 |
| 27.96 | 4,7,7-trimethylbycyclo(3.3.0)-octan-2-one | 1659 | ni | ni | ni | ni | ni | ni | 0.02 (0.01) | 1.670 | 0.132 |
| **Norisoprenoids and quinones** | | | | | | | | | | | |
| 20.41 | 3,5,5-trimethyl-2-Cyclohexen-1-one (alpha-Isophorone) | 1139 | ni | ni | 0.01 (0.02) | ni | 0.02 (0.07) | ni | ni | 2.500 | 0.025 |
| 20.84 | 2,6,6-trimethyl-Cyclohex-2-ene-1,4-dione (4-Ketoisophorone) | 1160 | ni | ni | ni | 0.03 (0.04) | 0.25 (0.38) | ni | ni | 8.373 | 0.000 |
| 20.98 | 2-hydroxy-3,5,5-trimethyl-2-Cyclohex-2-enone (2-hydroxylsophorone) | 1167 | ni | ni | 0.01 (0.01) | 0.08 (0.05) | 0.12 (0.20) | ni | ni | 7.102 | 0.000 |
| 22.83 | 2-methyl-5-propan-2-ylcyclohexa-2,5-diene-1,4-dione (Thymoquinone) | 1250 | ni | ni | ni | ni | ni | ni | 0.51 (1.25) | 3.178 | 0.006 |
| **Terpenoids** | | | | | | | | | | | |
| 16.18 | 2,6,6-trimethylbicyclo[3.1.1]Hept-2-ene (α-Pinene) | 948 | ni | ni | 0.04 (0.02) | 0.04 (0.02) | 0.09 (0.24) | 0.11 (0.54) | 0.06 (0.02) | 0.681 | 0.665 |
| 17.57 | Herboxide second isomer | 1005 | ni | 0.06 (0.08) | 0.02 (0.04) | ni | ni | ni | 0.04 (0.02) | 10.175 | 0.000 |
| 17.97 | 1-methyl-4-propan-2-ylCyclohexa-1,3-diene (α-Terpinene) | 1026 | ni | ni | ni | ni | ni | ni | 0.03 (0.10) | 1.623 | 0.145 |
| 18.26 | 1-methyl-4-prop-1-en-2-ylCyclohexene (dl-Limonene) | 1031 | ni | ni | ni | ni | 0.02 (0.05) | ni | 0.04 (0.10) | 1.935 | 0.079 |
| 18.38 | 4-methylene-1-(1-methylethyl)bicyclo(3.1.0)-Hexane (Sabinene) | 1044 | ni | ni | ni | ni | ni | ni | 0.05 (0.17) | 1.645 | 0.139 |
| 18.43 | 1,8-Cineole (Eucalyptol) | 1047 | ni | ni | ni | ni | ni | ni | 0.58 (2.16) | 1.386 | 0.224 |
| 18.85 | 1-methyl-4-propan-2-ylcyclohexa-1,4-diene (γ-Terpinene) | 1056 | ni | ni | ni | ni | ni | ni | 0.29 (0.86) | 2.198 | 0.047 |
| 19.14 | cis-Linalool oxide | 1073 | ni | 0.05 (0.06) | 0.17 (0.29) | 0.06(0.07) | 0.11 (0.13) | ni | 0.03 (0.06) | 7.502 | 0.000 |
| 19.54 | 3,7-dimethyl-1,6-Octadien-3-ol (Linalool) | 1088 | ni | 0.05 (0.04) | ni | 0.30 (0.11) | 0.14 (0.25) | ni | 0.40 (1.17) | 1.789 | 0.105 |
| 19.59 | 1-methyl-4-propan-2-yl-Cyclohexa-1,3-diene (α-Terpinolene) | 1090 | ni | 0.001 (0.005) | 0.20 (0.18) | ni | ni | ni | 0.03 (0.09) | 10.265 | 0.000 |

[a] RT: Retention time. RIexp: Experimental retention index values using standard hydrocarbons naturally present in honey. ni: not identified. The non-identified volatile compounds were treated as zeros for chemometrics and not as missing values. F: Values of the F distribution. p: probability.

Table 1. Cont.

| RT [a] (min) | Volatile Compounds (mg/kg) | RIexp | Clover | Citrus | Chestnut | Eucalyptus | Fir | Pine | Thyme | F | p |
|---|---|---|---|---|---|---|---|---|---|---|---|
| 19.63 | Hotrienol | 1104 | ni | 0.01 (0.02) | ni | 0.10 (0.20) | ni | ni | 0.06 (0.14) | 3.323 | 0.004 |
| 21.54 | (1S-endo)-1,7,7-trimethyl-bicyclo(2.2.1)-Heptan-2-ol (Borneol) | 1169 | ni | ni | 0.02 (0.03) | ni | ni | ni | 0.19 (0.48) | 2.860 | 0.012 |
| 21.61 | 4-methyl-1-propan-2-yfCyclohex-3-en-1-ol (Terpinen-4-ol) | 1178 | ni | ni | ni | ni | ni | ni | 0.39 (0.90) | 3.519 | 0.003 |
| 21.84 | 2(4-methylcyclohex-3-en-1-yl)Propan-2-ol (α-Terpineol) | 1191 | ni | 0.004 (0.02) | ni | ni | ni | ni | 0.04 (0.17) | 0.838 | 0.543 |
| 22.25 | *alpha*,4-dimethyl-cyclohex-3-ene-1-Acetaldehyde (*para*-Menth-1-en-9-al) | 1231 | ni | 0.04 (0.05) | ni | ni | ni | ni | 0.04 (0.02) | 11.818 | 0.000 |
| 26.32 | 1R,4E,9S)-4,11,11-Trimethyl-8-methylidenebicyclo(7.2.0)-undec-4-ene (Caryophyllene) | 1418 | ni | ni | ni | ni | ni | ni | 0.03 (0.12) | 1.243 | 0.288 |
| **Phenolic and benzene derivatives** | | | | | | | | | | | |
| 18.24 | 1-methyl-4-(1-methylethyl)Benzene (*para*-Cymene) | 1038 | ni | 0.03 (0.04) | ni | 0.03 (0.04) | ni | ni | 1.84 (5.00) | 2.528 | 0.023 |
| 20.64 | Benzeneacetonitrile | 1150 | ni | ni | ni | ni | ni | ni | 0.03 (0.07) | 2.799 | 0.013 |
| 23.31 | 5-methyl-2-(1-methylethyl)Phenol (Thymol) | 1296 | ni | 0.006 (0.003) | ni | ni | ni | ni | 3.51 (8.63) | 3.133 | 0.006 |
| 23.58 | 3-methyl-4-isopropylPhenol | 1308 | ni | ni | ni | ni | ni | ni | 0.003 (0.01) | 1.335 | 0.245 |
| 23.68 | 2-methyl-5-(1-methylethyl)Phenol (Carvacrol) | 1309 | ni | 0.01 (0.04) | ni | ni | ni | ni | 0.001 (0.002) | 0.777 | 0.590 |
| 24.11 | 3,4,5-trimethylPhenol | 1331 | ni | ni | 0.10 (0.17) | ni | 0.10 (0.25) | ni | ni | 3.609 | 0.002 |
| 24.73 | 2-methoxy-4-(2-propenyl)Phenol (Eugenol) | 1359 | ni | ni | ni | ni | ni | ni | 0.12 (0.32) | 2.428 | 0.029 |

[a] RT: Retention time. RIexp: Experimental retention index values using standard hydrocarbons naturally present in honey. ni: not identified. The non-identified volatile compounds were treated as zeros for chemometrics and not as missing values. F: Values of the F distribution. p: probability.

Pine honeys had higher amounts (mg/kg) of acetic acid, octane, *alpha*-pinene and *beta*-thujone. It is remarkable that *beta*-thujone was identified only in pine honeys, comprising a key volatile marker, even though the *p*-value ($p = 0.053$) was slightly higher than the level of confidence considered in the study. This finding is in agreement with a previous study on the botanical differentiation of citrus, fir, pine, and thyme honeys [8]. *Beta*-thujone has a menthol odor [24].

In the case of thyme honeys, 1-octen-3-ol, thymol methyl ether, carvacrol methyl ether, octyl ether, 4,7,7-trimethylbyciclo (3.3.0)-octan-2-one, and eugenol were found in higher amounts (mg/kg) compared to the other honey types. Ethers give an ether-like odor, whereas 1-octen-3-ol provides a floral and grassy odor. Furthermore, 1-Octen-3-ol is referred to as "mushroom alcohol", given the fact that is the main flavor component of mushrooms [25]. Eugenol is a phenylpropanoid volatile that has a pleasant, spicy, and clove-like odor [26].

### 3.2. Classification of Clover, Citrus, Chestnut, Eucalyptus, Fir, Pine, and Thyme Honeys According to Botanical Origin and the Honey Code Using Volatile Compounds in Combination with Chemometrics

#### 3.2.1. Part I. Classification of Honeys According to Botanical Origin

MANOVA and LDA

The four qualitative criteria of the multivariate hypothesis, namely Pillai's Trace = 5.573 ($F = 8.698$, df = 540, $p = 0.000$, $\eta^2 = 0.929$), Wilks' Lambda = 0.000 ($F = 22.251$, df = 540, $p = 0.000$, $\eta^2 = 0.972$), Hotelling's Trace = 1034.279 ($F = 102.151$, df = 540, $p = 0.000$, $\eta^2 = 0.994$), and Roy's Largest Root = 789.550 ($F = 526.367$, df = 90, $p = 0.000$, $\eta^2 = 0.999$) showed that there was a statistically significant effect of the botanical origin of honey on the semi-quantitative data of volatile compounds.

More specifically, 62 of the 94 volatile compounds were found to be significant ($p < 0.05$) for the botanical origin differentiation of honeys. Afterwards, these volatiles were subjected to LDA. The minimum tolerance level of the analysis was set at 0.001. Results showed that 4-terpineol, borneol, *para*-cymene, carvacrol methyl ether, thymoquinone, thymol, and eugenol did not pass the tolerance test. Therefore, these volatile compounds were excluded (SPSS program) a priori from the discriminant analysis. Therefore, 56 volatile compounds were subjected to LDA.

Results showed that six canonical discriminant functions were formed: Wilks' Lambda = 0.000, $X^2 = 1624.974$, df = 336, $p = 0.000$ for the first function; Wilks' Lambda = 0.000, $X^2 = 1049.108$, df = 275, $p = 0.000$ for the second function; Wilks' Lambda = 0.006, $X^2 = 609.334$, df = 216, $p = 0.000$ for the third function; Wilks' Lambda = 0.032, $X^2 = 406.682$, df = 159, $p = 0.000$ for the fourth function; Wilks' Lambda = 0.122, $X^2 = 249.608$, df = 104, $p = 0.000$ for the fifth function; and Wilks' Lambda = 0.409, $X^2 = 106,005$, df = 51, $p = 0.000$ for the sixth function.

The first discriminant function recorded the higher eigenvalue (127.977) and a canonical correlation of 0.996, accounting for 71.5% of total variance. The second discriminant function recorded a much lower eigenvalue (39.902) and a canonical correlation of 0.988, accounting for 22.3% of total variance. The third discriminant function recorded a much lower eigenvalue (4.530) and a canonical correlation of 0.905, accounting for 2.5% of total variance. Similarly, the fourth discriminant function recorded an even lower eigenvalue (2.764) and a canonical correlation of 0.857, accounting for 1.5% of total variance. Moreover, the fifth discriminant function had a lower eigenvalue (2.360) and a canonical correlation of 0.838, accounting for 1.3% of total variance. Finally, the sixth discriminant function had the lowest eigenvalue (1.446) and a canonical correlation of 0.769, accounting for 0.8% of total variance. All six discriminant functions accounted for 100% of total variance.

Figure 2 shows the differentiation of honeys according to botanical origin based on the use of 56 volatile compounds and LDA. The group centroid values which represent the unstandardized canonical discriminant functions evaluated at group means are also plotted. Each centroid gives information about the coordinates (discriminant functions) of the group means in the polyparametric space. The abscissa is the first discriminant function, and the ordinate is the second. So, the respective group centroid values were as follows: (−6.609, −0.891), (−6.081, −1.951), (−12.106, 39.255), (−5.664, 13.355), (21.632, 0.974), (−5.265, −1.395), and (−4.712, −2.268) for clover, citrus, chestnut, eucalyptus, fir, pine, and thyme honeys.

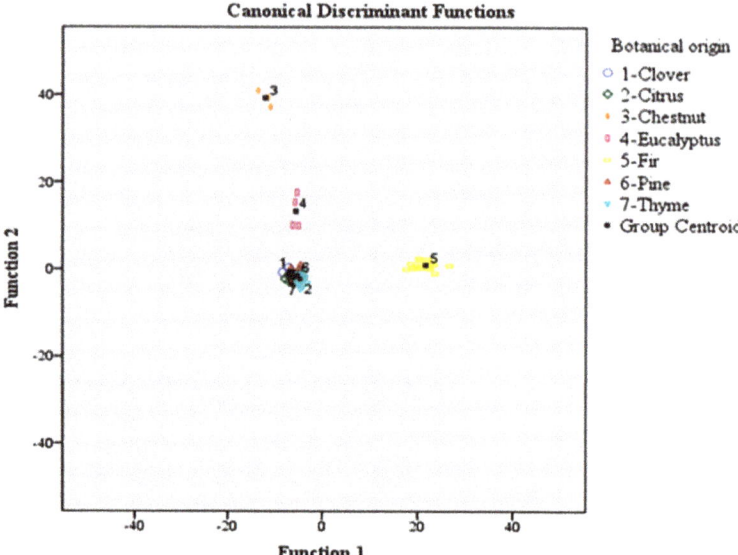

**Figure 2.** Classification of the 151 honey samples according to botanical origin based on the 56 volatile compounds and LDA. LDA: linear discriminant analysis.

The classification rate was 95.4% using the original and 81.5% using the cross-validation method. Supplementary Table S1 shows the significant variables (markers of botanical origin of clover, citrus, chestnut, eucalyptus, fir, pine, and thyme honeys) that were ordered by absolute size of correlation within function. The higher the absolute value of correlation, the best discrimination the variable provides within the discriminant function. The most pronounced markers of the botanical origin of honey are marked with an asterisk. In particular, octanoic acid ethyl ester, nonanoic acid ethyl ester, decanoic acid ethyl ester, dodecanoic acid ethyl ester, decanal, nonanal, 5-methyl-4-nonene, hexanoic acid ethyl ester, heptanoic acid ethyl ester, 4-ketoisophorone, octanol, tetradecanoic acid ethyl ester, geranyl acetone, 6-methyl-5-hepten-2-one, 1-(2-furanyl)-ethanone, *alpha*-isophorone, and decanol contributed to discriminant function 1, whereas hexadecanoic acid ethyl ester contributed to discriminant function 2. It should be remembered that the first two discriminant functions explained 93.8% of total variance.

The botanical origin classification rate of honeys, based on the original method, followed the sequence: clover (87.5%), citrus (90.3%), chestnut (100%), eucalyptus (100%), fir (100%), pine (100%), and thyme (91.4%). In total, 12.5% of clover honeys (one sample) were allocated to pine honeys. In total, 9.7% of citrus honeys (three samples) were allocated to pine honeys. Finally, 8.6% of thyme honeys (three samples) were allocated to pine honeys.

On the contrary, the botanical origin classification rate of honeys, based on the cross-validation method followed the sequence: clover (62.5%), citrus (80.6%), chestnut (66.7%), eucalyptus (75%), fir (100%), pine (94.9%), and thyme (57.1%). In total, 25% of clover honeys (two samples) were allocated to pine honeys, whereas 12.5% of samples (one sample) were allocated to thyme honeys. In total, 19.4% of citrus honeys (six samples) were allocated to pine honeys. In total, 33.3% of chestnut honeys (one sample) were allocated to fir honeys. In total, 25% of eucalyptus honeys (one sample) were allocated to chestnut honeys. In total, 5.2% of pine honeys (two samples) were allocated in equal proportions

(2.6%) to clover and chestnut honeys, respectively. Finally, 42.8% of thyme honeys were allocated to clover (11.4%) (four samples), citrus (2.9%) (one sample), chestnut (11.4%) (four samples), and pine (17.1%) (six samples) honeys, respectively (Table 2).

**Table 2.** Discriminatory power of the LDA model based on the significant volatile compounds according to the botanical origin of honey.

| Chemometric Technique LDA | Prediction Rate % | Botanical Origin | Predicted Group Membership | | | | | | | Total Honey Samples |
|---|---|---|---|---|---|---|---|---|---|---|
| | | | Clover | Citrus | Chestnut | Eucalyptus | Fir | Pine | Thyme | |
| Original [a] | Count | Clover | 7 | 0 | 0 | 0 | 0 | 1 | 0 | 8 |
| | | Citrus | 0 | 28 | 0 | 0 | 0 | 3 | 0 | 31 |
| | | Chestnut | 0 | 0 | 3 | 0 | 0 | 0 | 0 | 3 |
| | | Eucalyptus | 0 | 0 | 0 | 4 | 0 | 0 | 0 | 4 |
| | | Fir | 0 | 0 | 0 | 0 | 31 | 0 | 0 | 31 |
| | | Pine | 0 | 0 | 0 | 0 | 0 | 39 | 0 | 39 |
| | | Thyme | 0 | 0 | 0 | 0 | 0 | 3 | 32 | 35 |
| | % | Clover | 87.5 | 0.0 | 0.0 | 0.0 | 0.0 | 12.5 | 0.0 | 100.0 |
| | | Citrus | 0.0 | 90.3 | 0.0 | 0.0 | 0.0 | 9.7 | 0.0 | 100.0 |
| | | Chestnut | 0.0 | 0.0 | 100.0 | 0.0 | 0.0 | 0.0 | 0.0 | 100.0 |
| | | Eucalyptus | 0.0 | 0.0 | 0.0 | 100.0 | 0.0 | 0.0 | 0.0 | 100.0 |
| | | Fir | 0.0 | 0.0 | 0.0 | 0.0 | 100.0 | 0.0 | 0.0 | 100.0 |
| | | Pine | 0.0 | 0.0 | 0.0 | 0.0 | 0.0 | 100.0 | 0.0 | 100.0 |
| | | Thyme | 0.0 | 0.0 | 0.0 | 0.0 | 0.0 | 8.6 | 91.4 | 100.0 |
| Cross validated [b,c] | Count | Clover | 5 | 0 | 0 | 0 | 0 | 2 | 1 | 8 |
| | | Citrus | 0 | 25 | 0 | 0 | 0 | 6 | 0 | 31 |
| | | Chestnut | 0 | 0 | 2 | 0 | 1 | 0 | 0 | 3 |
| | | Eucalyptus | 0 | 0 | 1 | 3 | 0 | 0 | 0 | 4 |
| | | Fir | 0 | 0 | 0 | 0 | 31 | 0 | 0 | 31 |
| | | Pine | 1 | 0 | 0 | 0 | 1 | 37 | 0 | 39 |
| | | Thyme | 4 | 1 | 4 | 0 | 0 | 6 | 20 | 35 |
| | % | Clover | 62.5 | 0.0 | 0.0 | 0.0 | 0.0 | 25.0 | 12.5 | 100.0 |
| | | Citrus | 0.0 | 80.6 | 0.0 | 0.0 | 0.0 | 19.4 | 0.0 | 100.0 |
| | | Chestnut | 0.0 | 0.0 | 66.7 | 0.0 | 33.3 | 0.0 | 0.0 | 100.0 |
| | | Eucalyptus | 0.0 | 0.0 | 25.0 | 75.0 | 0.0 | 0.0 | 0.0 | 100.0 |
| | | Fir | 0.0 | 0.0 | 0.0 | 0.0 | 100.0 | 0.0 | 0.0 | 100.0 |
| | | Pine | 2.6 | 0.0 | 0.0 | 0.0 | 2.6 | 94.9 | 0.0 | 100.0 |
| | | Thyme | 11.4 | 2.9 | 11.4 | 0.0 | 0.0 | 17.1 | 57.1 | 100.0 |

[a] 95.4% of grouped cases using the original method were correctly classified. [b] Cross validation was performed only for those cases in the analysis. In cross validation, each case is classified by the functions derived from all cases rather than that particular case. [c] 81.5% of cross-validated grouped cases were correctly classified.

It should be stressed that these results were accepted, given the fact that cross validation is a more pessimistic method of classification of a group of objects, since in cross validation, each case is classified by the functions derived from all cases rather than that particular case. The errors obtained in both classification techniques (original and cross-validation) reveal important findings regarding honey authentication. These errors show the contribution of numerous plants in the produced honeys, even in cases of honeydew honeys. Honeydew honeys are harvested after nectar honeys. Therefore, the contribution of flowering plants in honeydew honeys is quite common.

In addition, present findings may be related to beekeeper practices during the harvesting of different honey types. However, the classification results obtained support previous studies in the literature that focus on honey authentication using volatile compound analysis and chemometrics [3–5,8,11]. The results obtained in the present study, which is collective in nature, show a clear differentiation of honeydew vs. nectar honeys.

K-Nearest Neighbors

For the k-NN analysis, the number of samples was randomly assigned to training and holdout partitions. The training sample consisted of 110 honey samples (72.8%), while the holdout sample consisted of 41 samples (27.2%). All cases (151 honey samples) (100%) were used in the statistical analysis, comprising a valid procedure.

The overall classification rate was 77.3% for the training and 87.8% for the holdout sample and was satisfactory in both cases. The botanical origin classification rate of honey types for the training sample followed the sequence: clover (75%), citrus (70.8%), chestnut (0%), eucalyptus (0%), fir (95.8%), pine (92.3%), and thyme (58.3%). However, the zero classification rates of chestnut and eucalyptus honeys are attributed to the limited honey samples, since the majority of them were assigned to the holdout sample. Of the eight clover honey samples subjected to training analysis, six were assigned to clover and two to thyme honeys. Similarly, of the 24 citrus honey samples, 17 samples were assigned to citrus, one to clover, four to pine, and two to thyme honeys, respectively. One chestnut honey sample was assigned to eucalyptus honeys. Of the 24 fir honey samples, 23 were assigned to fir and only one to pine honeys. A similar trend was obtained for pine honeys—in which, 25 samples were assigned to pine and only one to fir honeys, respectively. Finally, of the 24 thyme honey samples, 14 were assigned to thyme, six to pine, three to clover, and one to citrus honeys, respectively.

Regarding the holdout sample classification rate, this was higher than the training sample by 10.5%. The botanical origin classification rate of honey types for the holdout sample followed the sequence: citrus (85.7%), chestnut (0%), eucalyptus (100%), fir (100%), pine (100%), and thyme (81.8%). The clover honeys were assigned previously to training sample. Therefore, no classification results were obtained. Of the seven citrus honey samples subjected to holdout analysis, six were assigned to citrus and one to pine honeys. Of the two chestnut honeys, one was assigned to citrus and one to eucalyptus honeys, respectively. Finally, of the 11 thyme honey samples, nine samples were assigned to thyme and two to pine, respectively (Table 3).

**Table 3.** Classification of clover, citrus, chestnut, eucalyptus, fir, pine, and thyme honeys according to botanical origin using the 56 volatile compounds and K-Nearest Neighbors (k-NN) analysis.

| Partition | Observed | Predicted | | | | | | | Percent Correct |
|---|---|---|---|---|---|---|---|---|---|
| | | Clover | Citrus | Chestnut | Eucalyptus | Fir | Pine | Thyme | |
| Training | Clover | 6 | 0 | 0 | 0 | 0 | 0 | 2 | 75.0% |
| | Citrus | 1 | 17 | 0 | 0 | 0 | 4 | 2 | 70.8% |
| | Chestnut | 0 | 0 | 0 | 1 | 0 | 0 | 0 | 0.0% |
| | Eucalyptus | 0 | 0 | 0 | 0 | 0 | 0 | 3 | 0.0% |
| | Fir | 0 | 0 | 0 | 0 | 23 | 1 | 0 | 95.8% |
| | Pine | 0 | 0 | 0 | 0 | 1 | 25 | 0 | 96.2% |
| | Thyme | 3 | 1 | 0 | 0 | 0 | 6 | 14 | 58.3% |
| | Overall Percent | 9.1% | 16.4% | 0.0% | 0.9% | 21.8% | 32.7% | 19.1% | 77.3% |
| Holdout | Clover | 0 | 0 | 0 | 0 | 0 | 0 | 0 | |
| | Citrus | 0 | 6 | 0 | 0 | 0 | 1 | 0 | 85.7% |
| | Chestnut | 0 | 1 | 0 | 1 | 0 | 0 | 0 | 0.0% |
| | Eucalyptus | 0 | 0 | 0 | 1 | 0 | 0 | 0 | 100.0% |
| | Fir | 0 | 0 | 0 | 0 | 7 | 0 | 0 | 100.0% |
| | Pine | 0 | 0 | 0 | 0 | 0 | 13 | 0 | 100.0% |
| | Thyme | 0 | 0 | 0 | 0 | 0 | 2 | 9 | 81.8% |
| | Missing | 0 | 0 | 0 | 0 | 0 | 0 | 0 | |
| | Overall Percent | 0.0% | 17.1% | 0.0% | 4.9% | 17.1% | 39.0% | 22.0% | 87.8% |

Among the 55 significant volatile compounds (predictors), the most effective predictors (k-nearest neighbors) that built the model were acetic acid ethyl ester, formic acid ethyl ester, and 2-methylbutanal. Based on this observation, the k-NN analysis was run again by performing feature selection in order to investigate whether the classification results could be improved. The selected features were the aforementioned volatile compounds.

For the specified k-nearest neighbors analysis, the sample was divided again to training and holdout partitions. The training sample consisted of 100 honey samples (66.2%), whereas the holdout sample consisted of 51 (33.8%). Both partitions explained 100% of the procedure, indicating that all cases were valid. The analysis stopped when the absolute ratio was less than or equal to the minimum change, which was inserted by default equal to 0.01. The overall classification rate was 83% for the training and 74.5% for the holdout sample. The individual botanical classification rate was differentiated given the fact that the sample assignment to training and holdout partitions was also differentiated (Table 4).

**Table 4.** Classification of clover, citrus, chestnut, eucalyptus, fir, pine, and thyme honeys according to botanical origin using the 10 volatile compounds and k-NN.

| Partition | Observed | Predicted | | | | | | | Percent Correct |
|---|---|---|---|---|---|---|---|---|---|
| | | Clover | Citrus | Chestnut | Eucalyptus | Fir | Pine | Thyme | |
| Training | Clover | 5 | 1 | 0 | 0 | 0 | 1 | 0 | 71.4% |
| | Citrus | 0 | 21 | 0 | 0 | 0 | 0 | 1 | 95.5% |
| | Chestnut | 0 | 0 | 0 | 0 | 1 | 1 | 1 | 0.0% |
| | Eucalyptus | 0 | 1 | 0 | 0 | 0 | 1 | 1 | 0.0% |
| | Fir | 0 | 0 | 0 | 0 | 17 | 0 | 0 | 100.0% |
| | Pine | 0 | 0 | 0 | 0 | 1 | 25 | 0 | 96.2% |
| | Thyme | 1 | 4 | 0 | 0 | 0 | 2 | 15 | 68.2% |
| | Overall Percent | 6.0% | 27.0% | 0.0% | 0.0% | 19.0% | 30.0% | 18.0% | 83.0% |
| Holdout | Clover | 0 | 1 | 0 | 0 | 0 | 0 | 0 | 0.0% |
| | Citrus | 0 | 5 | 0 | 0 | 0 | 4 | 0 | 55.6% |
| | Chestnut | 0 | 0 | 0 | 0 | 0 | 0 | 0 | |
| | Eucalyptus | 0 | 0 | 0 | 0 | 0 | 1 | 0 | 0.0% |
| | Fir | 0 | 0 | 0 | 0 | 14 | 0 | 0 | 100.0% |
| | Pine | 0 | 2 | 0 | 0 | 0 | 10 | 1 | 76.9% |
| | Thyme | 0 | 1 | 0 | 0 | 0 | 3 | 9 | 69.2% |
| | Missing | 0 | 0 | 0 | 0 | 0 | 0 | 0 | |
| | Overall Percent | 0.0% | 17.6% | 0.0% | 0.0% | 27.5% | 35.3% | 19.6% | 74.5% |

Considering the total standard error of the forced predictors, but also that of the individual predictors, the model built with $k = 3$ (nearest neighbors) was more applicable. The 10 predictors were the three specified predictors, followed by furfural, lilac aldehyde C, benzaldehyde, nonanol, *para*-cymene, 5-methyl-4-nonene, and nonane. The total error rate of the three specified features was 0.74. The respective individual error rate for the seven predictors left was: 0.39, 0.27, 0.25, 0.21, 0.18, 0.17, and 0.17.

On the contrary, the total error rate of the model built with $k = 4$ nearest neighbors in relation to the three specified features was 0.75. The individual error rate for the seven predictors left was: 0.41, 0.28, 0.25, 0.23, 0.21, 0.20, and 0.20 for furfural, lilac aldehyde C, phenylethylalcohol, 5-methyl-4-nonene, 1-octen-3-ol, and nonane, respectively. The selection of the predictors during $k$-NN analysis with $k = 3$ and $k = 4$ nearest neighbors according to the botanical origin of honey is shown in Figure 3.

FA

During factor analysis, the 56 significant volatile compounds were reduced to 16 principal components (PCs) based on the rule of an eigenvalue greater than one. The rotated component matrix is given in Supplementary Table S2. The total variance explained was 80.524% (ca. 80.52%).

The fitness of data for factor analysis was estimated by the Kaiser–Meyer–Olkin (KMO) test, which comprises a measure of the effective performance of factor analysis, to a set of data, indicating the sampling adequacy. The acceptable value was considered that of KMO ≥ 0.50. The suitability and applicability of factor analysis was further evaluated using Bartlett's test of sphericity. This test highlights the hypothesis that the correlation matrix is an identity matrix, which would indicate that the variables incorporated into the model are unrelated and therefore unsuitable for structure detection. Small probability values ($p < 0.05$) indicate that a factor analysis may be useful with data treatment [20]. The value of the KMO test was 0.636. Furthermore, Bartlett's test of sphericity had the following qualitative values: $X^2 = 10{,}587.564$, df = 1540, and $p = 0.000$.

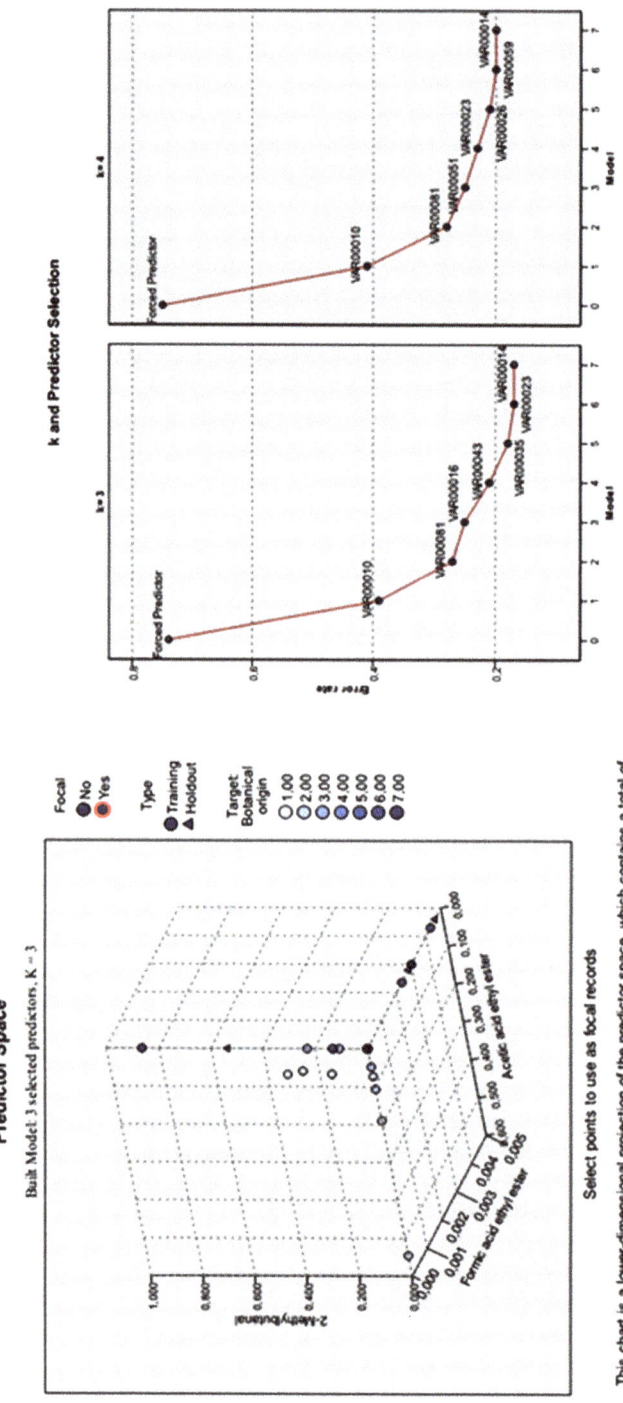

**Figure 3.** Predictor selection during $k$-NN analysis with $k = 3$ and $k = 4$ nearest neighbors. 1: Clover honeys. 2: Citrus honeys. 3: Chestnut honeys. 4: Eucalyptus honeys. 5: Fir honeys. 6: Pine honeys. 7: Thyme honeys. Forced predictor: Acetic acid ethyl ester, formic acid ethyl ester, 2-methylbutanal. VAR00010: Furfural. VAR00081: Lilac aldehyde C. VAR00016: Benzaldehyde. VAR00043: Nonanol. VAR00035: *para*-Cymene. VAR00023: 5-methyl-4-Nonene. VAR00014: Nonane (Model with $k = 3$). VAR00010: Furfural. VAR00081: Lilac aldehyde C. VAR00051: Phenylethylalcohol. VAR00023: 5-methyl-4-Nonene. VAR00026: 1-Octen-3-ol. VAR00059: Decanol.VAR00014: Nonane (Model with $k = 4$).

The factors that best explained the rotated component matrix were the following volatile compounds: Decanol (PC1, 10.454% of total variance), undecanoic acid ethyl ester (PC2, 10.380% of total variance), *para*-cymene (PC3, 7.768% of total variance), 2-hydroxyisophorone (PC4, 7.699% of total variance), nonane (PC5, 5.509% of total variance), dill ether (PC6, 5.058% of total variance), lilac aldehyde C (PC7, 4.412% of total variance), lilac aldehyde D (PC8, 4.352% of total variance), acetic acid ethyl ester (PC9, 3.918% of total variance), 5-methyl-2-phenylhexenal (PC10, 3.905% of total variance), decane (PC11, 3.596% of total variance), 1-(2-furanyl)-ethanone (PC12, 3.563% of total variance), benzeneacetonitrile (PC13, 3.240% of total variance), nonanol (PC14, 2.400% of total variance), 2-methylbutanal (PC15, 2.156% of total variance), and hexanoic acid ethyl ester (PC16, 2.113% of total variance).

3.2.2. Part II. Classification of Honeys According to the Honey Code (CCC-E-F-P-T)

MANOVA and LDA

As in the case of the botanical origin differentiation of honeys, the four qualitative criteria of the multivariate hypothesis had the following values: Pillai's Trace = 3.770 ($F$ = 10.640, df = 364, $p$ = 0.000, $\eta^2$ = 0.943), Wilks' Lambda = 0.000 ($F$ = 23.673, df = 364, $p$ = 0.000, $\eta^2$ = 0.974), Hotelling's Trace = 431.692 ($F$ = 64.635, df = 364, $p$ = 0.000, $\eta2$ = 0.991), and Roy's Largest Root = 361.350 ($F$ = 234.282, df = 91, $p$ = 0.000, $\eta^2$ = 0.997) showed that there was a statistically significant effect of the honey code on the semi-quantitative data of volatile compounds. More specifically, 65 of the 94 volatile compounds were found to be significant ($p < 0.05$) for the classification of honeys according to the honey code. Then, these volatiles were subjected to LDA. Results showed that 4-terpineol, (1R)-1,7,7-trimethyl-bicyclo-(2.2.1)-heptan-2-one, carvacrol methyl ether, thymoquinone, thymol, and eugenol did not pass the tolerance test. Therefore, these volatile compounds were excluded from the discriminant analysis. In that sense, 59 volatile compounds contributed to the structure matrix as shown in Supplementary Table S3.

Results showed that four canonical discriminant functions were formed: Wilks' Lambda = 0.000, $X^2$ = 1009.601, df = 236, $p$ = *0.000* for the first function; Wilks' Lambda = 0.014, $X^2$ = 502.352, df = 174, $p$ = *0.000* for the second function; Wilks' Lambda = 0.090, $X^2$ = 284.776, df = 114, $p$ = *0.000* for the third function; and Wilks' Lambda = 0.369, $X^2$ = 117.597, df = 56, $p$ = 0.000 for the fourth function. The first discriminant function recorded the higher eigenvalue (72.605) and a canonical correlation of 0.993, accounting for 87.7% of total variance. The second discriminant function recorded a much lower eigenvalue (5.321) and a canonical correlation of 0.917, accounting for 6.4% of total variance. The third discriminant function recorded a lower eigenvalue (3.124) and a canonical correlation of 0.870, accounting for 3.8% of total variance. Similarly, the fourth discriminant function recorded the lowest eigenvalue (1.709) and a canonical correlation of 0.794, accounting for 2.1% of total variance. All four discriminant functions accounted for 100% of total variance.

In Figure 4, the differentiation of the 151 honey samples according to the honey code is shown. The group centroid values are as follows: (−4.968, −1.828), (−2.691, −4.647), (16.454, −0.115), (−4.012, −0.914), (−3.834, 3.844) for honeys encoded with CCC, E, F, P, and T, respectively.

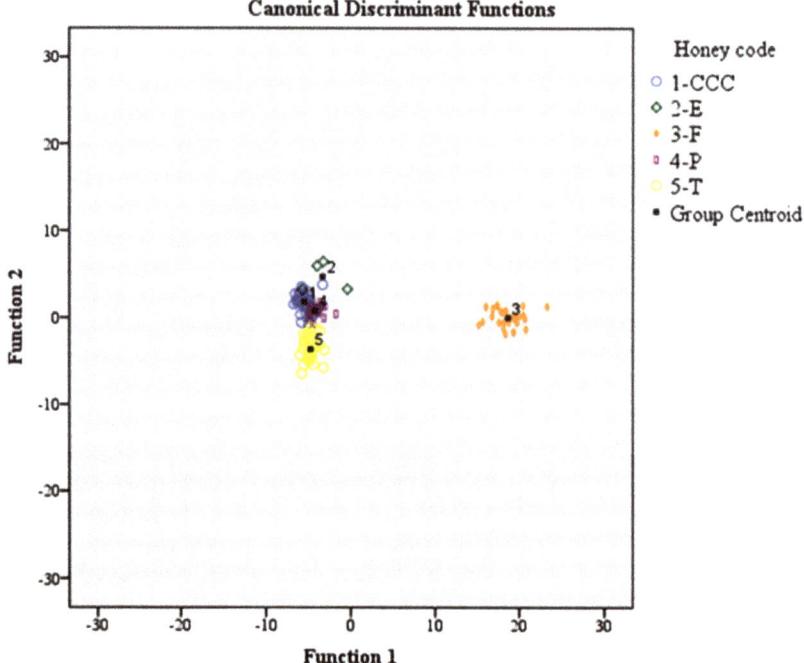

**Figure 4.** Classification of the 151 honey samples according to the honey code based on 57 volatile compounds and LDA.

The classification rate was 96.0% using the original and 86.1% using the cross-validation method. The classification rate of honeys according to the honey code, based on the original method, followed the sequence: CCC (88.1%), E (100%), F (100%), P (100%), and T (97.1%). In total, 12.9% of CCC honeys were allocated to the E group (4.8%, two samples) and to the P group (7.1%, three samples). In total, 2.9% (three samples) of T honeys were allocated to the P group.

For the cross-validation method, the classification rate of honeys followed the sequence: CCC (81%), E (75%), F (93.5%), P (92.3%), and T (80%). In total, 4.8% (two samples) of CCC honeys were allocated to the E group; 2.4% (one sample) were allocated to the F group, and 11.9% of samples (five samples) were allocated to the P group. In total, 25% (one sample) of E honeys were allocated to CCC honeys. In total, 3.2% of F honeys were allocated in equal proportions to the E (one sample) and P (one sample) groups, respectively. The same holds for P honeys—in which, two samples (5.2% of the total population) were allocated to the E (one sample, 2,6%) and F (one sample, 2,6%) groups, respectively. Finally, 14.3% of T honeys were allocated to P honeys (five samples), and 5.7% (two samples) to E honeys (Table 5).

As can be observed, the classification of honeys according to the honey code based on original and cross validation methods provided higher prediction rates compared to the differentiation of honeys according to botanical origin. In Table 3, the key volatile compounds for the discrimination of honeys according to the honey code are marked with an asterisk.

**Table 5.** Discriminatory power of the LDA model based on the significant volatile compound according to the honey code.

| Chemometric Technique | Prediction Rate | Honey Code | Predicted Group Membership | | | | | Total Honey Samples |
|---|---|---|---|---|---|---|---|---|
| LDA | % | | CCC | E | F | P | T | (n = 151) |
| Original [a] | Count | CCC | 38 | 2 | 0 | 2 | 0 | 42 |
| | | E | 0 | 4 | 0 | 0 | 0 | 4 |
| | | F | 0 | 0 | 31 | 0 | 0 | 31 |
| | | P | 0 | 0 | 0 | 39 | 0 | 39 |
| | | T | 0 | 0 | 0 | 1 | 34 | 35 |
| | % | CCC | 90.5 | 4.8 | 0.0 | 4.8 | 0.0 | 100.0 |
| | | E | 0.0 | 100.0 | 0.0 | 0.0 | 0.0 | 100.0 |
| | | F | 0.0 | 0.0 | 100.0 | 0.0 | 0.0 | 100.0 |
| | | P | 0.0 | 0.0 | 0.0 | 100.0 | 0.0 | 100.0 |
| | | T | 0.0 | 0.0 | 0.0 | 2.9 | 97.1 | 100.0 |
| Cross validated [b,c] | Count | CCC | 35 | 2 | 1 | 4 | 0 | 42 |
| | | E | 1 | 3 | 0 | 0 | 0 | 4 |
| | | F | 0 | 1 | 29 | 1 | 0 | 31 |
| | | P | 0 | 0 | 1 | 37 | 1 | 39 |
| | | T | 0 | 2 | 0 | 4 | 29 | 35 |
| | % | CCC | 83.3 | 4.8 | 2.4 | 9.5 | 0.0 | 100.0 |
| | | E | 25.0 | 75.0 | 0.0 | 0.0 | 0.0 | 100.0 |
| | | F | 0.0 | 3.2 | 93.5 | 3.2 | 0.0 | 100.0 |
| | | P | 0.0 | 0.0 | 2.6 | 94.9 | 2.6 | 100.0 |
| | | T | 0.0 | 5.7 | 0.0 | 11.4 | 82.9 | 100.0 |

[a] 96.7% of grouped cases using the original method were correctly classified. [b] Cross validation was performed only for those cases in the analysis. In cross validation, each case is classified by the functions derived from all cases rather than that particular case. [c] 88.1% of cross validated grouped cases were correctly classified.

### K-Nearest Neighbors

The training sample consisted of 111 honey samples that represented 73.5% of the total sample population. Similarly, the holdout sample consisted of 40 samples that represented 26.5% of the sample population. All cases (151 honey samples) (100%) were used in the statistical analysis, comprising a valid procedure. The scale features (predictors) (57 statistically significant volatile compounds) were normalized during analysis.

The overall classification rate was 81.1% for the training and 80% for the holdout sample and was satisfactory in both cases. The classification rate of honey types according to the honey code followed the sequence: CCC (86.7%), E (25%), F (96.2%), P (96.4%), and T (47.8%). Of the 30 CCC samples subjected to training analysis, one sample was assigned to the E group and three samples to the P group. Similarly, of the four E honey samples, three honey samples were assigned to the CCC group and only one sample to the E group. Of the 26 F honeys, 25 samples were assigned to the F group and only one sample to the P group. Of the 28 P honeys, 27 samples were assigned to the P group and only one sample to the F group. Finally, of the 23 T honey samples, 11 were assigned to T, six to CCC, and six to P, respectively. For the training sample, the honey code had the following hierarchy: CCC > E, F, P > T, CCC > T, and was in general applicable.

The classification rates for the holdout sample were improved for the F, P, and T honey groups. The classification rate of honey samples according to the honey code for the holdout sample followed the sequence: CCC (83.3%), F (100%), P (100%), and T (50%). As can be observed, the hierarchy in honey lettering was maintained by 3/5 cases (F, P > T and CCC > T). Of the 12 CCC honey samples assigned to the holdout sample, 10 were assigned to the CCC group and two to the P group. F and P honey samples were perfectly assigned to the respective groups. Finally, of the 12 T honey samples, seven were assigned to T group, two to the CCC, and four to the P group (Table 6).

**Table 6.** Classification of clover, citrus, chestnut, eucalyptus, fir, pine, and thyme honeys according to the honey code using the 57 volatile compounds and *k*-NN analysis.

| Partition | Observed | Predicted | | | | | Percent Correct |
|---|---|---|---|---|---|---|---|
| | | CCC | E | F | P | T | |
| Training | CCC | 26 | 1 | 0 | 3 | 0 | 86.7% |
| | E | 3 | 1 | 0 | 0 | 0 | 25.0% |
| | F | 0 | 0 | 25 | 1 | 0 | 96.2% |
| | P | 0 | 0 | 1 | 27 | 0 | 96.4% |
| | T | 6 | 0 | 0 | 6 | 11 | 47.8% |
| | Overall Percent | 31.5% | 1.8% | 23.4% | 33.3% | 9.9% | 81.1% |
| Holdout | CCC | 10 | 0 | 0 | 2 | 0 | 83.3% |
| | E | 0 | 0 | 0 | 0 | 0 | |
| | F | 0 | 0 | 5 | 0 | 0 | 100.0% |
| | P | 0 | 0 | 0 | 11 | 0 | 100.0% |
| | T | 2 | 0 | 0 | 4 | 6 | 50.0% |
| | Missing | 0 | 0 | 0 | 0 | 0 | |
| | Overall Percent | 30.0% | 0.0% | 12.5% | 42.5% | 15.0% | 80.0% |

Among the 57 significant volatile compounds (predictors), the most effective predictors (*k*-nearest neighbors) that built the model were formic acid ethyl ester, acetic acid ethyl ester, and heptane. Therefore, the *k*-NN analysis was repeated by performing feature selection in order to investigate whether the classification results could be improved. The selected features were the aforementioned volatile compounds.

For the specified *k*-nearest neighbors, the sample was divided again to the training and holdout partitions. The training sample consisted of 110 honey samples (72.8%), whereas the rest of the honey samples represented the holdout sample (27.2%). Similar to the botanical origin differentiation of honeys, the analysis stopped when the absolute ratio was less than or equal to the minimum change which was inserted by default equal to 0.01.

The overall classification rate was 89.1% for the training and 63.4% for the holdout sample. The classification rate of the training sample was improved, whereas that of the holdout sample was decreased. However, discrepancies among the two methods followed (simple *k*-NN and *k*-NN analysis with feature selection) may be attributed to the number of the predictors assigned to the model and the random dividing of the sample to training and holdout partitions in relation to sample size.

The classification rate of honey samples according to the honey code for the training sample followed the sequence: CCC (87.5%), E (0%), F (100%), P (93.1%), and T (88.5%). These results show that the honey code was satisfactorily applicable given the hierarchy followed: CCC > E, F > P > T and CCC > T. The classification rate of honey samples according to the honey code for the holdout sample followed the sequence: CCC (50%), E (0%), F (72.7%), P (80%), and T (55.6%). Similarly, the honey code was satisfactorily applicable given the hierarchy followed: CCC > E, F, P > T, and CCC > T. The classification results along with each sample assignment are given in Table 7.

The model was built with $k = 3$, $k = 4$, and $k = 5$ nearest neighbors, which were automatically selected. Similar to the botanical origin differentiation of honeys, the forced predictors' error was considered for the selection of the best model. The 10 predictors that were obtained during the *k*-NN analysis with $k = 3$ neighbors were the three forced predictors (heptane, formic acid ethyl ester, acetic acid ethyl ester) followed by dodecanoic acid ethyl ester, benzaldehyde, nonanal, *alpha*-terpinene, geranyl acetone, 5-methyl-4-nonene, and lilac aldehyde B (isomer II). The total error rate of the three specified features was 0.6818. The respective individual error rate for the seven aforementioned predictors was: 0.30, 0.2727, 0.2455, 0.1818, 0.1616, 0.1455, 0.1364, and 0.1364.

On the other hand, the total error rate of the model built with $k = 4$ nearest neighbors in relation to the three specified features was 0.6636. The individual error rate for the six predictors left was: 0.3000,

0.1909, 0.1636, 0.1364, 0.1273, and 0.1273, for furfural, lilac aldehyde C (isomer III), decanoic acid, lilac aldehyde D (isomer IV), pentanoic acid, and nonane.

Finally, the total error rate of the model built with $k = 5$ nearest neighbors in relation to the three specified features was 0.6545. The individual error rate for the seven predictors left was: 0.2909, 0.20, 0.1364, 0.1545, 0.1182, 0.1091, and 0.1091, for furfural, lilac aldehyde C (isomer III), benzeneacetaldehyde, octanoic acid ethyl ester, nonanoic acid ethyl ester, nonanoic acid, and pentanoic acid. The selection of the predictors during $k$-NN analysis with $k = 3$, $k = 4$, and $k = 5$ nearest neighbors according to the honey code is shown in Figure 5.

**Table 7.** Classification of clover, citrus, chestnut, eucalyptus, fir, pine, and thyme honeys according to the honey code using the forced predictors and, in total, 10 volatile compounds and $k$-NN analysis.

| Partition | Observed | Predicted | | | | | Percent Correct |
|---|---|---|---|---|---|---|---|
| | | CCC | E | F | P | T | |
| Training | CCC | 28 | 0 | 0 | 2 | 2 | 87.5% |
| | E | 2 | 0 | 0 | 0 | 1 | 0.0% |
| | F | 0 | 0 | 20 | 0 | 0 | 100.0% |
| | P | 0 | 0 | 1 | 27 | 1 | 93.1% |
| | T | 2 | 0 | 0 | 1 | 23 | 88.5% |
| | Overall Percent | 29.1% | 0.0% | 19.1% | 27.3% | 24.5% | 89.1% |
| Holdout | CCC | 5 | 0 | 1 | 3 | 1 | 50.0% |
| | E | 0 | 0 | 0 | 0 | 1 | 0.0% |
| | F | 0 | 0 | 8 | 3 | 0 | 72.7% |
| | P | 0 | 0 | 1 | 8 | 1 | 80.0% |
| | T | 2 | 0 | 0 | 2 | 5 | 55.6% |
| | Missing | 0 | 0 | 0 | 0 | 0 | |
| | Overall Percent | 17.1% | 0.0% | 24.4% | 39.0% | 19.5% | 63.4% |

FA

The 59 significant volatile compounds were reduced to 15 principal components (PCs) based on the rule of an eigenvalue greater than one. The rotated component matrix is given in Supplementary Table S4. The total variance explained was 81.547% (ca. 81.55%). As can be observed, the total variance explained of samples according to the honey code was higher than those that were grouped according to botanical origin. In addition, the KMO test had a higher value: KMO = 0.615. Furthermore, Bartlett's test of sphericity had the following qualitative values: $X^2 = 12,376.576$, df = 1596, $p = 0.000$. The factors that best explained the rotated component matrix were the following volatile compounds: *Para*-cymene (PC1, 12.207% of total variance), undecanoic acid ethyl ester (PC2, 10.520% of total variance), decanol (PC3, 10.03% of total variance), 2-hydroxyisophorone (PC4, 7.788% of total variance), decanoic acid (PC5, 7.386% of total variance), dill ether (PC6, 5.078% of total variance), 1-(2-furanyl)-ethanone (PC7, 4.506% of total variance), undecane (PC8, 4.326% of total variance), lilac aldehyde B (PC9, 3.254% of total variance), dodecanoic acid (PC10, 2.989% of total variance), lilac aldehyde D (PC11, 2.986% of total variance), benzeneacetonitrile (PC12, 2.904% of total variance), decanoic acid ethyl ester (PC13, 2.821% of total variance), 4,7,7-trimethylbyciclo (3.3.0)-octan-2-one (PC14, 2.819% of total variance), and dodecanoic acid ethyl ester (PC15, 1.960% of total variance).

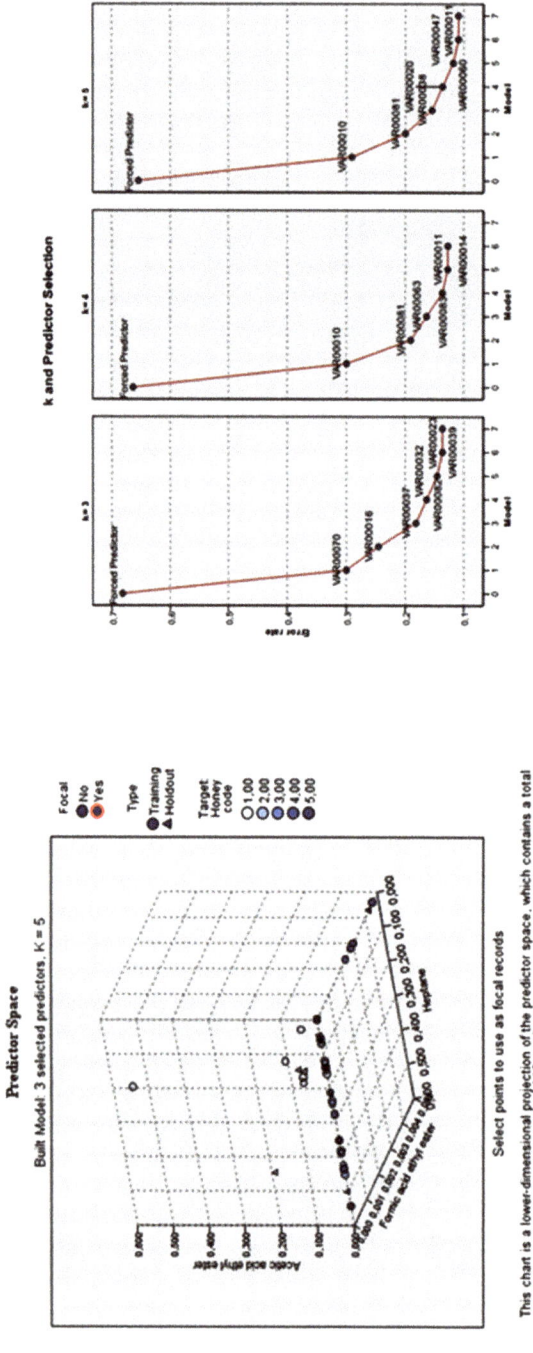

**Figure 5.** Predictor selection during *k*-NN analysis with $k = 3$, $k = 4$, and $k = 5$ nearest neighbors. 1: CCC. 2: E. 3: F. 4: P. 5: T. Forced predictor: Heptane, formic acid ethyl ester, acetic acid ethyl ester. VAR00070: Dodecanoic acid ethyl ester. VAR00016: Benzaldehyde. VAR00037: Nonanal. VAR00032: *Alpha*-terpinene. VAR00082: Geranyl acetone. VAR00023: 5-Methyl-4-Nonene. VAR00039: Lilac aldehyde B (isomer II). (Model with $k = 3$); VAR00010: Furfural. VAR00081: Lilac aldehyde C (isomer III). VAR00063: Decanoic acid. VAR00083: Lilac aldehyde D (isomer IV). VAR00011: Pentanoic acid. VAR00014: Nonane (Model with $k = 4$); VAR00010: Furfural. VAR00081: Lilac aldehyde C (isomer III). VAR00020: Benzeneacetaldehyde. VAR00038: Octanoic acid ethyl ester. VAR00060: Nonanoic acid ethyl ester. VAR00047: Nonanoic acid. VAR00011: Pentanoic acid (Model with $k = 5$).

## 4. Conclusions

Results of the present study showed that specific volatile compounds in combination with polyparametric statistical techniques such as MANOVA, LDA, $k$-NN and FA, may provide exhaustive information about the botanical origin of honey. Even though honey samples were harvested in different parts of the world, the classification of honeys according to botanical origin was, in general, very satisfactory. At the same time, the application of the honey code to the collective set of data and the use of the aforementioned statistical techniques resulted in a higher classification rate of the honey samples. The use of hierarchical classification strategies (HCSs) may expand the state of the art and flourish the complicated topic of "Honey authentication" by highlighting with numbers the distinction/differentiation rates, for instance, of monofloral or blends of nectar or honeydew honeys with specific and intense flavors.

**Supplementary Materials:** The following are available online at http://www.mdpi.com/2304-8158/8/10/508/s1, Table S1: Standardized canonical discriminant function coefficients of the discrimination model—structure matrix of the LDA model according to the botanical origin of honey; Table S2: Rotated component matrix of volatile compounds used for the botanical origin differentiation of clover, citrus, chestnut, eucalyptus, fir, pine, and thyme honeys; Table S3: Standardized canonical discriminant function coefficients of the discrimination model—structure matrix of the LDA model according to the honey code; Table S4: Rotated component matrix of volatile compounds used for the distinction of clover, citrus, chestnut, eucalyptus, fir, pine, and thyme honeys according to the honey code; Figure S1: A typical gas chromatogram of citrus honey (no. 4) from Spain indicating selected key volatile compounds. 10: Herboxide (isomer II). 11: Lilac aldehyde C (isomer III). 12: Dill ether. IS: internal standard; Figure S2: A typical gas chromatogram of chestnut honey (no. 2) from Portugal indicating selected key volatile compounds. 13: Octane. 14: 5-methyl-2-Furaldehyde. 15: Benzaldehyde. 16: Benzeneacetaldehyde. IS: internal standard; Figure S3: A typical gas chromatogram of eucalyptus honey (no. 1) from Portugal indicating selected key volatile compounds. 17: Heptane. 18: beta-Damascenone IS: internal standard; Figure S4: A typical gas chromatogram of fir honey (no. 6) from Greece indicating selected key volatile compounds. 19: Nonane. 20: 1-(2-furanyl)-Ethanone. 21: 6-methyl-5-Hepten-2-one. 22: 5-methyl-4-Nonene. 23: Hexanoic acid ethyl ester. 24: Heptanoic acid ethyl ester. 25: Nonanal. 26: *alpha*-Isophorone. 27: 4-Ketoisophorone. 28: 2-Hydroxyisophorone. 29: Octanoic acid ethyl ester. 30: Decane. 31: Nonanoic acid ethyl ester. 32: Geranyl acetone. IS: internal standard; Figure S5: A typical gas chromatogram of pine honey (no. 2) from Greece indicating selected key volatile compounds. 33: Acetic acid. 34: Octane. 35: *alpha*-Pinene. 36: *beta*-Thujone. IS: internal standard; Figure S6: A typical gas chromatogram of thyme pine (no. 6) from Egypt indicating selected key volatile compounds. 37: *alpha*-Pinene. 38: I-Octen-3-ol. 39: *alpha*-Terpinene. 40: *para*-Cymene. 41: Camphor. 42: Carvacrol methyl ether. 43: Thymoquinone. 44: 4,7,7-trimethylbyciclo(3.3.0)-Octan-2-one. IS: internal standard.

**Author Contributions:** Conceptualization, I.K.K.; methodology, I.K.K. and A.V.B.; software, A.V.B and I.K.K. validation, I.K.K.; formal analysis, I.K.K. and V.K.K.; investigation, I.K.K. and V.K.K.; resources, A.V.B.; data curation, I.K.K. and V.K.K.; writing—original draft preparation, I.K.K.; writing—review and editing, I.K.K.; visualization, I.K.K.; supervision, I.K.K..; project administration, I.K.K.; funding acquisition, A.V.B.

**Funding:** This research received no external funding.

**Acknowledgments:** The authors are grateful to Attiki Bee Culturing Co., Alex. Pittas S.A, Protomagias 9, Kryoneri 14568, Athens, Greece, and professional beekeepers from the studied regions for the donation of honey samples. Miguel Maia (APISMAIA) is greatly acknowledged for the collection of Portuguese honey samples. Michael G. Kontominas is greatly acknowledged for the collection of honey samples from Egypt, Morocco and Spain.

**Conflicts of Interest:** The authors declare no conflict of interest.

## References

1. Crane, E. *The World History of Beekeeping and Honey Hunting*, 1st ed.; Routledge: Abingdon, UK, 1999; 704p.
2. *Council Directive 2001/110/EC Relating to Honey*; Council of the European Union: Brussels, Belgium, 2002; pp. 47–52.
3. Castro-Vázquez, L.; Díaz-Maroto, M.C.; González-Viñas, M.A.; Perez-Coello, M.S. Aroma composition and new chemical markers of Spanish citrus honeys. *Food Chem.* **2007**, *103*, 601–606. [CrossRef]
4. Kadar, M.; Juan-Borrás, M.; Carot, M.J.; Domenech, E.; Escriche, I. Volatile fraction composition and physicochemical parameters as tools for the differentiation of lemon blossom honey and orange blossom honey. *J. Sci. Food Agric.* **2011**, *91*, 2768–2776. [CrossRef] [PubMed]

5. Nayik, G.A.; Nanda, V. Characterization of the volatile profile of unifloral honey from Kashmir valley of India by using solid-phase microextraction and gas chromatography-mass spectrometry. *Eur. Food Res. Technol.* **2015**, *240*, 1091–1100. [CrossRef]
6. Piasenzotto, L.; Gracco, L.; Conte, L. SPME applied to honey quality control. *J. Sci. Food Agric.* **2003**, *83*, 1037–1044. [CrossRef]
7. Manyi-Loh, C.E.; Ndip, R.N.; Clarke, A.M. Volatile compounds in honey: A review on their involvement in aroma, botanical origin determination and potential biomedical activities. *Int. J. Mol. Sci.* **2011**, *12*, 9514–9532. [CrossRef] [PubMed]
8. Karabagias, I.K.; Badeka, A.V.; Kontakos, S.; Karabournioti, S.; Kontominas, M.G. Botanical discrimination of Greek unifloral honeys with physico-chemical and chemometric analyses. *Food Chem.* **2014**, *165*, 181–190. [CrossRef] [PubMed]
9. Bianchi, F.; Careri, M.; Musci, M. Volatile norisoprenoids as markers of botanical origin of Sardinian strawberry-tree (*Arbutus unedo* L.) honey: Characterisation of aroma compounds by dynamic headspace extraction and gas chromatography–mass spectrometry. *Food Chem.* **2005**, *89*, 527–532. [CrossRef]
10. Karabagias, I.K.; Nikolaou, C.; Karabagias, V.K. Volatile fingerprints of common and rare honeys produced in Greece: In Search of PHVMs with implementation of the honey code. *Eur. Food Res. Technol.* **2019**, *245*, 23–39. [CrossRef]
11. Castro-Vázquez, L.; Díaz-Maroto, M.C.; González-Viñas, M.A.; Perez-Coello, M.S. Differentiation of monofloral citrus, rosemary, eucalyptus, lavender, thyme and heather honeys based on volatile composition and sensory descriptive analysis. *Food Chem.* **2009**, *112*, 1022–1030. [CrossRef]
12. Escriche, I.; Visquert, M.; Juan-Borras, M.; Fito, P. Influence of simulated industrial thermal treatments on the volatile fractions of different varieties of honey. *Food Chem.* **2009**, *112*, 329–338. [CrossRef]
13. Bayraktar, D.; Onoğur, T.A. Investigation of the aroma impact volatiles in Turkish pine honey samples produced in Marmaris, Datça and Fethiye regions by SPME/GC–MS technique. *Int. J. Food Sci. Technol.* **2011**, *46*, 1060–1065. [CrossRef]
14. Petretto, G.L.; Tuberoso, C.I.G.; Vlahopoulou, G.; Atzei, A.; Mannu, A.; Zrira, S.; Pintore, G. Volatiles, color characteristics and other physico-chemical parameters of commercial Moroccan honeys. *Nat. Prod. Res.* **2016**, *30*, 286–292. [CrossRef] [PubMed]
15. Moreira, R.F.A.; De Maria, C.A.B. Investigation of the aroma compounds from headspace and aqueous solution from the cambara (*Gochnatia Velutina*) honey. *Flavour Frag. J.* **2005**, *20*, 13–17. [CrossRef]
16. Karabagias, I.K.; Louppis, P.A.; Karabournioti, S.; Kontakos, S.; Papastephanou, C.; Kontominas, M.G. Characterization and classification of commercial thyme honeys produced in specific Mediterranean countries according to geographical origin, using physicochemical parameter values and mineral content in combination with chemometrics. *Eur. Food Res. Technol.* **2017**, *243*, 889–900. [CrossRef]
17. Patrignani, M.; Fagundez, G.A.; Tananaki, C.; Thrasyvoulou, A.; Lupano, C.E. Volatile compounds of Argentinean honeys: Correlation with floral and geographical origin. *Food Chem.* **2018**, *246*, 32–40. [CrossRef] [PubMed]
18. Miller, J.N.; Miller, J.C. *Statistics and Chemometrics for Analytical Chemistry*, 6th ed.; Pearson Education Limited: London, UK, 2010.
19. Krishnamoorthy, K. *Statistical Tolerance Regions: Theory, Applications, and Computation*; John Wiley and Sons: Hoboken, NJ, USA, 2009; pp. 1–6.
20. International Business Machines (IBM). IBM Knowledge Center. 2019. Available online: https://www.ibm.com/support/knowledgecenter/ (accessed on 11 August 2019).
21. Cserháti, T.; Forgács, E. *Encyclopedia of Food Sciences and Nutrition*, 2nd ed.; Flavor (Flavour) Compounds, Structures and Characteristics; Elsevier Science Ltd.: Amsterdam, The Netherlands, 2003; pp. 2509–2517.
22. Karabagias, I.K.; Halatsi, E.Z.; Kontakos, S.; Karabournioti, S.; Kontominas, M.G. Volatile fraction of commercial thyme honeys produced in Mediterranean regions and key volatile compounds for geographical discrimination: A chemometric approach. *Int. J. Food Prop.* **2017**, *20*, 2699–2710. [CrossRef]
23. Hefetz, A.; Eickwort, G.C.; Blum, M.S.; Cane, J.; Bohart, G.E.J. A comparative study of the exocrine products of cleptoparasitic bees (*Holcopasites*) and their hosts (*Calliopsis*) (Hymenoptera: Anthophoridae, Andrenidae). *Chem. Ecol.* **1982**, *8*, 1389–1397. [CrossRef] [PubMed]

24. Perry, N.B.; Anderson, R.E.; Brennan, N.J.; Douglas, M.H.; Heaney, A.J.; McGimpsey, J.A.; Smallfield, B.M. Essential oils from Dalmatian sage (*Salvia officinalis* L.): Variations among individuals, plant parts, seasons, and sites. *J. Agric. Food Chem.* **1999**, *47*, 2048–2054. [CrossRef] [PubMed]
25. "1-Octen-3-ol". Available online: http://www.thegoodscentscompany.com (accessed on 31 May 2015).
26. Human Metabolome Database: Showing Metabocard for Eugenol (HMDB0005809). Available online: www.hmdb.ca (accessed on 1 July 2018).

© 2019 by the authors. Licensee MDPI, Basel, Switzerland. This article is an open access article distributed under the terms and conditions of the Creative Commons Attribution (CC BY) license (http://creativecommons.org/licenses/by/4.0/).

Article

# HPTLC-PCA Complementary to HRMS-PCA in the Case Study of *Arbutus unedo* Antioxidant Phenolic Profiling

Mariateresa Maldini [1,2,*], Gilda D'Urso [3], Giordana Pagliuca [4], Giacomo Luigi Petretto [1], Marzia Foddai [1], Francesca Romana Gallo [4], Giuseppina Multari [4], Donatella Caruso [2], Paola Montoro [3] and Giorgio Pintore [1]

1. Department of Chemistry and Pharmacy, University of Sassari, Via F. Muroni, 23/b, 07100 Sassari, Italy
2. Department of Pharmacological and Biomolecular Sciences, Università degli Studi di Milano, Via Balzaretti, 9, 20133 Milan, Italy
3. Department of Pharmacy, University of Salerno, Via Giovanni Paolo II, 84084 Fisciano (SA), Italy
4. National Center for Drug Research and Evaluation, Viale Regina Elena, 299, 00161 Roma, Italy
* Correspondence: maldiniluce@gmail.com; Tel.: +39-02-5031-8323

Received: 20 June 2019; Accepted: 20 July 2019; Published: 27 July 2019

**Abstract:** A comparison between High-Performance Thin-Layer Chromatography (HPTLC) analysis and Liquid Chromatography High Resolution Mass Spectrometry (LC–HRMS), coupled with Principal Component Analysis (PCA) was carried out by performing a combined metabolomics study to discriminate *Arbutus unedo* (*A. unedo*) plants. For a rapid digital record of *A. unedo* extracts (leaves, yellow fruit, and red fruit collected in La Maddalena and Sassari, Sardinia), HPTLC was used. Data were then analysed by PCA with the results of the ability of this technique to discriminate samples. Similarly, extracts were acquired by non-targeted LC–HRMS followed by unsupervised PCA, and then by LC–HRMS (MS) to identify secondary metabolites involved in the differentiation of the samples. As a result, we demonstrated that HPTLC may be applied as a simple and reliable untargeted approach to rapidly discriminate extracts based on tissues and/or geographical origins, while LC–HRMS could be used to identify which metabolites are able to discriminate samples.

**Keywords:** HPTLC; LC–HRMS; PCA; metabolomics; Arbutus unedo; antioxidant activities

## 1. Introduction

Plant metabolic profiling is a very useful strategy to study the complexity and the large variety of compounds belonging to different chemical classes [1] and is ideally suited to comparing many samples in order to classify them according to botanical, geographical origins and chemotypes [2–5].

As a consequence, one of the many aims of research in the field of metabolomics is to analyze a large number of samples and obtain information in the shortest times and with a little or no sample preparation time [6,7]. Creation of rapid and convenient methods for simultaneous metabolites fingerprinting and their quantification requires the use of peculiar analytical techniques.

In the last few years, the progress and developments of analysis techniques, including advanced hyphenated techniques (like Liquid Chromatography–Mass Spectrometry (LC–MS) and Gas Chromatography–Mass Spectrometry (GC–MS)), can effectively satisfy this demand [8–10].

Until now, most metabolomics studies were performed using the most popular analytical technologies, such as Nuclear Magnetic Resonance, GC, LC, and MS [11]. The new developments reached, employing high chromatographic resolution/separation interfaced with high resolution mass spectrometry, showed the power of this coupled technique in metabolomics [12–15]. High Performance Thin Layer Chromatography (HPTLC) is gaining more attraction in the field of metabolomics. In fact, recently, several papers have reported how HPTLC is an alternative method for routine analysis of complex matrices [5,16–18].

If, on one side, liquid chromatography–high resolution mass spectrometry is almost the preferred technique to perform screening studies of complex matrices, on the other hand, it requires many steps for metabolomics studies, in primis, among all, time run acquisitions, processing and normalization of raw data (baseline and phase correction, alignment), and, in some cases, a long and expensive sample preparation, too. In this context, High Performance Thin Layer Chromatography (HPTLC) has been acknowledged as a potent and suited analytical platform because it shows high accuracy, adaptability, flexibility, and reproducibility and because it is able to generate a quick fingerprint of diverse samples in a unique analysis [19]. HPTLC fingerprinting coupled with Principal Component Analysis (PCA) results in a consistent untargeted approach that is able to distinguish samples based on the tissues and their geographical origin [20–22].

*Arbutus unedo*, also named "strawberry tree", is an evergreen shrub belonging to the Ericaceae family. It is distributed mainly in the Mediterranean region, in particular in Southern European countries, where it grows especially near the sea. Fruits and leaves possess a wide series of biological activities, such as astringent, depurative, diuretic, anti-inflammatory, antioxidant properties, attributable to their rich content in phenolic compounds. In addition, *Arbutus unedo* produces edible berries and covers economic importance in the sense of its use for the production of beverages, liquors, jams, marmalades, and of a unfloral honey. Sardinia is probably the largest producer in the world of this honey with a characteristic taste, also known as "bitter honey" [23,24].

The specific purposes of the present study were: (a) Compare the possibility to use HPTLC–PCA as an alternative to more comprehensive LC– Elettrospray (ESI)–Orbitrap–MS PCA approaches to preliminary discriminate samples of *A. unedo* (leaves, yellow fruit, and red fruit collected in Sassari and in the archipelago of La Maddalena). The samples were from two somewhat different habitats: Plants from Sassari grew about 15 km from the sea (220–230 m a.s.l.), while plants from La Maddalena grew in a coastal site (0–20 m a.s.l.); (b) identify secondary metabolites by LC–ESI–Orbitrap–MS and LC–ESI–Orbitrap–MS/MS and evaluate tentatively which ones contribute most to the distinction of the samples; (c) study the relationship between antioxidant activity assayed by DPPH (2,2-Diphenyl-1-picrylhydrazyl) and ABTS (2,2′-azino-bis(3-ethylbenzothiazoline-6-sulphonic acid)) and total phenolic content performed by Folin–Ciocalteu's method.

## 2. Materials and Methods

### 2.1. Reagents and Chemicals

HPLC–MS grade methanol, acetonitrile, and formic acid were purchased from Sigma–Aldrich Chemical Company (St Louis, MO, USA). HPLC grade water (18 mΩ) was obtained by using a Milli-Q purification system, Millipore (Bedford, MA, USA). Chemicals and reagents necessary for antioxidant activity assays were supplied by Sigma (Dorset, UK).

### 2.2. Sampling Sites and Extraction

Wild plants of *A. unedo* were collected in October 2015 in two selected geographical areas of Sardinia: Sassari and La Maddalena. Botanical identity of the plants was assigned by Doctor M. Chessa. Voucher specimens were left at the Erbarium SASSA of Sassari University (number 514).

Samples were collected in triplicate, thus resulting in 18 biological samples, classified into 6 groups as in the following: LMFG (Yellow Fruit La Maddalena); LMFR (Red Fruit La Maddalena); LML (Leaves La Maddalena); SSFG (Yellow Fruit Sassari); SSFR (Red Fruit Sassari); SSL (Leaves Sassari).

Leaves and yellow and red fruits of *A. unedo* from Sassari and La Maddalena were extracted under ultrasound agitation for 1 h, with ethanol/water (70:30 $v/v$ using a sample to solvent ratio 1:10 $w/v$); then samples were stored in the dark overnight. Samples were then filtered and dried using a rotary evaporator under a vacuum and temperature of 30 °C, working in the dark.

For qualitative analysis, dried samples were dissolved again in methanol to generate a solution of 1 mg/mL. The solutions were filtered through 0.20 µm syringe PVDF filters (Whatmann International Ltd., Maidstone, UK).

## 2.3. HPTLC Analyses

The HPTLC analyses were performed following the protocol described by Maldini et al., 2016 [16], with slight modification. Extracts were reconstituted at a concentration of 6 mg/mL and a volume of 6 µL was loaded. The developing solution for HPTLC plates was a mixture of ethyl acetate/dichloromethane/acetic acid/formic acid/water (100:25:10:10:11; $v/v/v/v/v$). The length of the chromatogram run was 70 mm from the point of application.

For the densitometric analysis, a CAMAG TLC scanner 3 (CAMAG, Muttenz, Switzerland) linked to winCATS software (version 1.2.1, CAMAG, Muttenz, Switzerland). was set at 254 nm and 366 nm, after an optimization performed by a multi-wavelength mode from 220 to 700 nm. A minimum background compensation was performed on the $x$-axis during the scanning. Deuterium and tungsten lamps were used as sources of radiation. The slit dimension was kept at $6.00 \times 0.45$ mm and the scanning speed was 100 mm/s.

## 2.4. LC–ESI–Orbitrap–MS Analysis

To characterize the main metabolites representative of each sample, the LC–ESI–Orbitrap–MS method was developed. Each extract was dissolved 1:100 with methanol and a 10 µL aliquot injected into the analytical system. A duplicate of each sample was carried out, obtaining a total of 36 analyzed samples. Experiments were run using a Thermo Scientific liquid chromatography system, equipped with a quaternary Accela 600 pump and an Accela auto sampler, in conjunction with a linear Trap–Orbitrap hybrid mass spectrometer (LTQ–Orbitrap XL, Thermo Fisher Scientific, Bremen, Germany), combining a linear trap quadrupole (LTQ) and an Orbitrap mass analyzer with an electrospray ionization (ESI) source. Chromatographic separation was obtained using an X-Select T3 C18 reversed phase column ($2.1 \times 150$ mm, 3.5 µm particle size) (Waters, Milford, Massachusetts). The mobile phase consisted of solvent A (water + 0.1% formic acid) and solvent B (acetonitrile + 0.1% formic acid). A linear gradient program at a flow rate of 0.200 mL/min was used: 0–3 min, from 0 to 10% (B); 3–25 min, from 10 to 20% (B); 25–35 min, from 20 to 30% (B); 35–40 min, from 30 to 60% (B); 40–45 min, from 60 to 100% (B); then 0% (B) for 5 min. The mass spectrometer was operated in negative ion mode. The ESI source parameters were the following: The capillary voltage −48 V; tube lens voltage −176.47 V; capillary temperature 300 °C; Sheath and Auxiliary Gas flow (N2) 15 and 5 (arbitrary units); Sweep gas 0 (arbitrary units); Spray voltage 3.50 V. MS spectra were acquired by full range acquisition covering $m/z$ 200–1200 (Resolution: 30,000). For MS/MS experiments, a data-dependent scan experiment was established, with the selection of precursor ions corresponding to the most intense peaks observed in the previous LC–MS analysis (threshold value 300).

Compounds were identified on the basis of their spectral characteristic fragmentation and retention time, with comparison with data reported in literature and databases. Data acquisition, data analysis, and instrument control were performed by Xcalibur software version 2.1 (Thermo Scientific™, Waltham, MA, USA)

## 2.5. PCA

Principal component analysis (PCA) is a multivariate data analysis technique used to reveal important patterns correlating to physiological, genetic, and environmental issues and has been used widely in assessing the differences between plant varieties at a metabolomics level [25]. An m × n matrix (where m is the number of samples and n is the number of variables) was used in PCA analysis of data obtained from HPTLC. For matrix building, the variables were taken from the pseudo-chromatogram and reported as the area % corresponding to intensities of the individual retention time factors of the most intense spots in the fingerprint of each sample. PCA was performed on the dataset scaled by unit variance with the Factor MineR package of R 2.15.2 software (R Foundation for Statistical Computing, Vienna, Austria). The results of the analysis are presented in terms of score and loading plots.

Similarly, an m × n matrix (where m is the number of samples and n is the number of variables) was used in PCA analysis of data obtained from LC–ESI–Orbitrap–MS. The untargeted approach was obtained working with the base peak chromatograms derived from LC–MS (negative ion mode), which were evaluated using a platform independent open source software package called MZmine (http://mzmine.sourceforge.net/). By means of this toolbox normalization by total raw signal and excluding noise from LC–MS profiles (Noise level 5.0 E3—all data points below this intensity level were ignored), 280 peaks were detected and the peak area was determined [26].

After transferring the processed data in tabular format (cvs file), further analysis of the data matrix (36 observation and 280 variables) were made by SIMCA (+) software 12.0 (Umetrix AB, Umea, Sweden) by PCA. PCA was achieved by measuring the peak areas obtained from LC/MS analysis [27]. Pareto scaling was applied to data before multivariate data analysis.

### 2.6. Antiradical Activity by Diphenyl-1-Picarylhydrazyl (DPPH) and TEAC Assays

The radical scavenging activity assay was performed according to the method proposed by Brand–Williams (1995) [28] with some modifications.

The ABTS free radical scavenging activity of each sample was determined according to the method described by Petretto et al. (2015) [29].

### 2.7. Determination of Total Phenols

Total phenols were estimated by a colorimetric assay based on procedures described by Lizcano et al. (2010) [30] and by Maldini et al., 2016 [16].

### 2.8. Statistical Analysis

All experiments were carried out in triplicate. Statistical analyses were performed by comparison of ethanolic extracts from red fruits, yellow fruits, and leaves of *Arbutus unedo* collected in different areas of Sardinia, with unpaired Student's *t*-test using Sigma-Stat v. 3.5 software (Systat Software GmbH, Erkrath, Germany). The distribution of samples was performed by the Kolmogorov–Smirnov and Shapiro tests. The strength of association between variables was investigated with the Pearson product moment correlation coefficient (data normally distributed). A $p \leq 0.05$ was considered statistically significant.

## 3. Results and Discussion

### 3.1. HPTLC–PCA Analysis

The power of HPTLC analysis is to get characteristic fingerprints of secondary metabolites occurring in numerous biological samples (~20) in a single run. Thus, it was here employed to quickly compare *A. unedo* samples (red fruit, yellow fruit, and leaves) from different places of collection.

By using an optimized separation method, the different pigments are highlighted, using UV light at 254 or 366 nm or under reflectance and transmission white light (WRT Figure 1A). HPTLC metabolomics patterns didn't show qualitative differences among samples collected in Sassari or in La Maddalena (Figure 1A). Successively, densitometric analysis was carried out at 254 nm and 366 nm. The HPTLC 3D chromatogram recorded at 254 nm in Figure 1B shows the characteristic traces of *A. unedo* red fruit, yellow fruit, and leaves from La Maddalena (1–9) and Sassari (10–18). Densitometer scanning at 254 nm wavelength was the most useful to characterize and differentiate extracts; thus, Rf values with a corresponding area for each spot were obtained and designated to produce an m × n matrix made up by 18 observations and 16 variables (metabolites detected). A PCA was conducted on the dataset to provide an overview to discriminate samples on the basis of the tissue and the geographical origin. Figure 2 shows the Score Plot obtained for a studied data set. The first principal component explains 68% of the variance, while the second principal component explains the 14%. Along the first component, samples are separated into two groups: In the negative values is located the group relative to leaves and in the positive values, red and yellow fruit. Along the second component, samples are located on the basis of their origin. All samples from La Maddalena are placed at positive values, whilst samples from Sassari are at negative values.

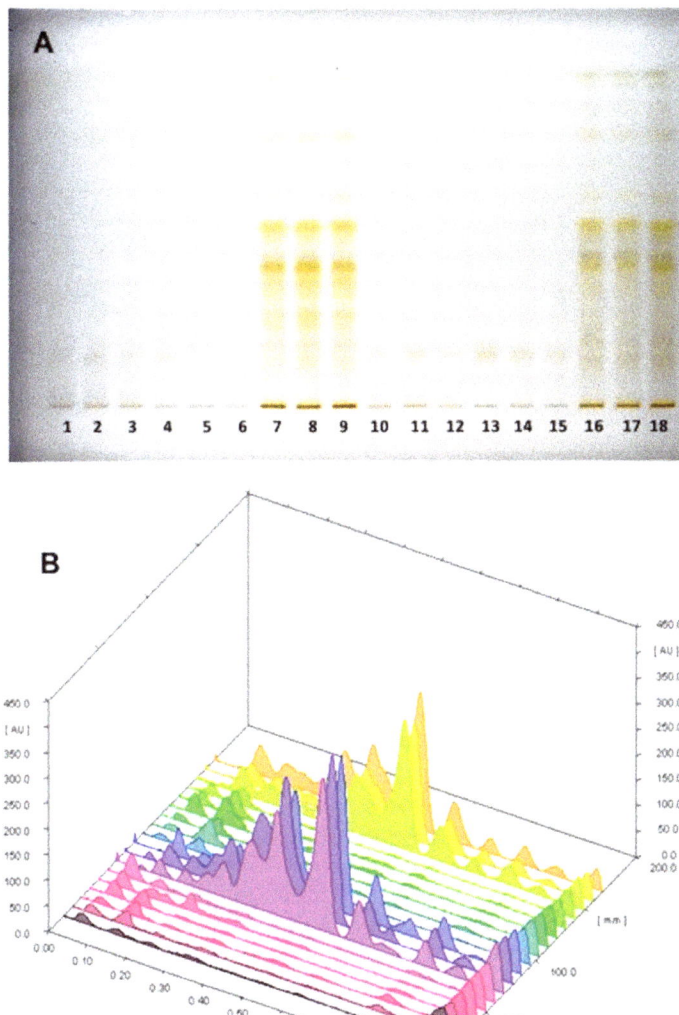

**Figure 1.** High Performance Thin Layer Chromatography (HPTLC) fingerprint (**A**) and three-dimensional (3D) densitograms (**B**) of *Arbutus unedo* samples (red fruit, yellow fruit, and leaves) from different places of collection (Sassari (SS) and archipelago of La Maddalena (LM)). 1. Red fruit LM 1; 2. Red fruit LM 2; 3. Red fruit LM 3; 4. Yellow fruit LM 1; 5. Yellow fruit LM 2; 6. Yellow fruit LM 3; 7. Leaves LM 1; 8. Leaves LM 2; 9. Leaves LM 3; 10. Red fruit SS 1; 11. Red fruit SS 2; 12. Red fruit SS 3; 13. Yellow fruit SS 1; 14. Yellow fruit SS 2; 15. Yellow fruit SS 3; 16. Leaves SS 1; 17. Leaves SS 2; 18. Leaves SS 3.

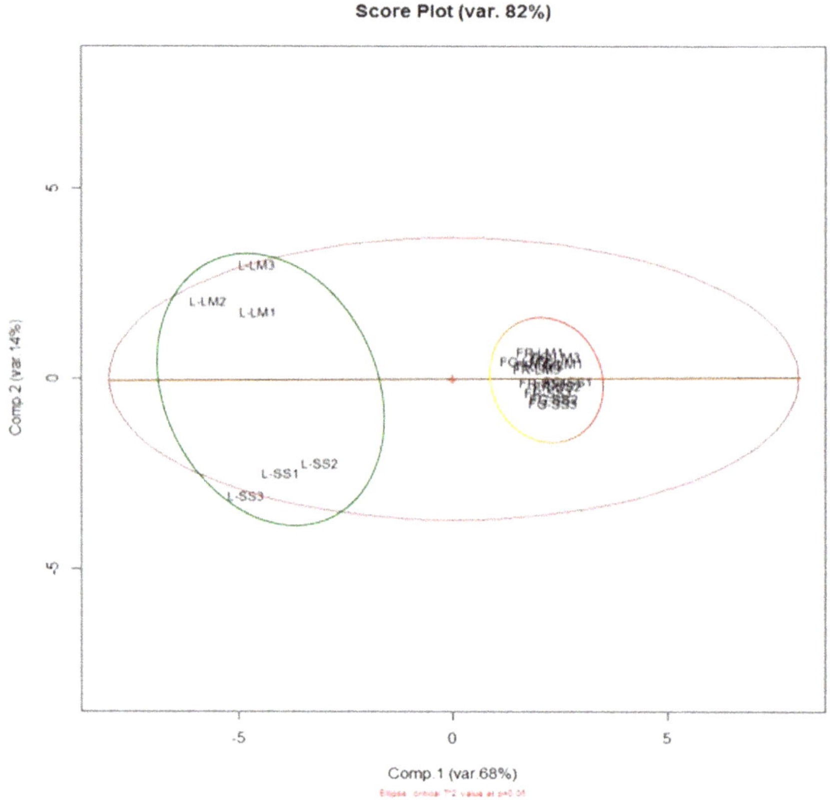

## Score Plot (var. 82%)

[Score plot showing Comp 1 (var 68%) on x-axis and Comp 2 (var 14%) on y-axis, with sample clusters labeled L-LM3, L-LM2, L-LM1, L-SS1, L-SS2, L-SS3, and FR/FG samples from LM and SS]

**Figure 2.** Score plot obtained from HPTLC data. (L = Leaves; FR = Red Fruit; FG = Yellow Fruit; SS = Sassari; LM = La Maddalena).

### 3.2. LC–ESI–Orbitrap–MS Analysis and PCA

With the aim to identify which secondary metabolites are responsible with a major impact for the variation of the analyzed samples, metabolites profiling of *A. unedo* leaves and red and yellow fruit ethanolic extracts from La Maddalena and Sassari were analyzed by liquid chromatography, coupled with high resolution mass spectrometry. Nineteen metabolites were detected and putatively identified (Table 1) according to the information obtained from accurate mass fragmentation spectra and then the obtained results were confirmed with those present in the literature and database. All the identified compounds were already reported in *A. unedo*. The main detected metabolites belong to flavonoids, principally quercetin, kaempferol, and myricetin derivatives (9–19) [31–35]. Quercetin and kaempferol derivatives were both similarly detected in leaves and fruit of *A. unedo* from Sassari and La Maddalena area, while myricetin derivatives are differently distributed in the samples; in fact, myricetin hexoside (9) was detected only in leaves. Other identified compounds belong to catechin and epicatechins (3 and 5) and galloyl organic acid derivative groups (2 and 8) [31] that, in the present study, were found both in fruit and leaves of the analyzed samples. In addition, other secondary metabolites belonging to ellagitannins class (4 and 7) were detected. These were found in all analysed samples, with the exception of yellow fruit of La Maddalena (LMFG) and red fruit of Sassari (SSFR). Finally, chlorogenic acid (6), present in all analysed samples, and arbutin (1), detected in all the samples except in yellow fruit from Sassari, were also identified. Figure 3 shows LC–HRMS and extracted ion chromatograms of compounds 1–19 identified in La Maddalena leaves extract.

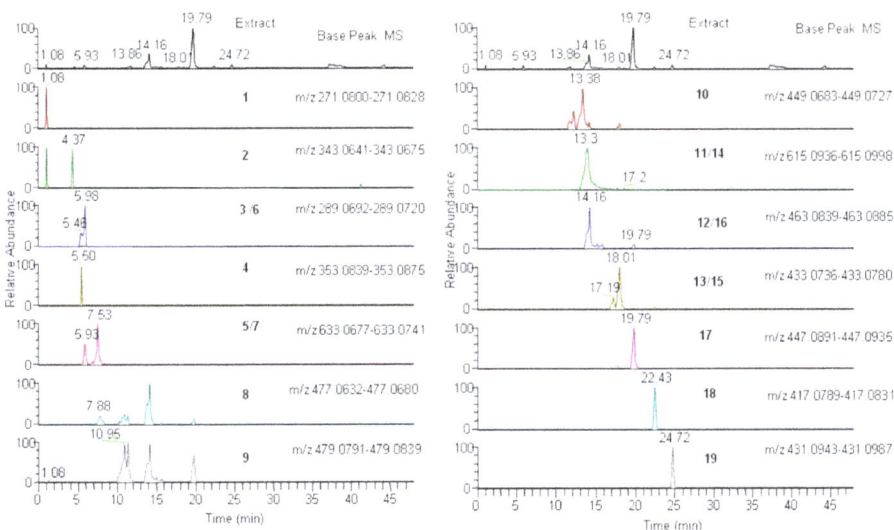

**Figure 3.** LC–ESI–MS/MS (Liquid Cromatography Elettrospray Ionization Mass (Mass) Spectrometry) profile and reconstructed ion chromatograms of *Arbutus* leaves extract. Number corresponds to the identified metabolites reported in Table 1.

The collected LC–ESI–Orbitrap–MS data were then processed by multivariate data analysis. With the aim to perform a comparative study of *Arbutus* extracts, the PCA approach was applied in an untargeted fashion to acquired raw data. LC/MS chromatograms were then pre-processed with the software MZmine to compensate for differences in retention time and $m/z$ between the chromatographic runs. The pre-processed chromatograms were transferred as a peak list table, with rows corresponding to the individual samples, and columns corresponding to the integrated and normalized peak areas.

PCA was developed by working with the peak areas of the total peaks present in the LC/MS dataset (excluding the noisy) in such a way that a matrix was obtained considering these areas, each corresponding to specific $m/z$ values (variables), and the column of the matrix was realized by the different analyzed samples. Score scatter and loading plots are reported in Figure 4A,B, respectively (LC–ESI–Orbitrap–MS data obtained in negative analysis). The first component explained the 50% of variance while the second explained the 8.9%. The legend of classification can be read as follows: LMFG (Yellow Fruit La Maddalena); LMFR (Red Fruit La Maddalena); LML (Leaves La Maddalena); SSFG (Yellow Fruit Sassari); SSFR (Red Fruit Sassari); SSL (Leaves Sassari). The score plot shows that HR–LC–MS–PCA discriminates samples on the basis of the tissue and of their geographical area, and confirms data obtained in previous analysis, apart from yellow and red fruit where geographical discrimination is less clear than that obtained by HPTLC–PCA.

Table 1. Metabolites identified in *A. unedo* leaves and fruit extracts by LC–ESI/LTQOrbitrap/MS and LC–ESI/LTQOrbitrap/MS/MS analysis. (x = present; nd = not detected).

| No. | RT | [M − H]− | Molecular Formula | Δ PPM | MS/MS | Identity | LMFG | LMFR | LML | SSFG | SSFR | SSL |
|---|---|---|---|---|---|---|---|---|---|---|---|---|
| 1 | 1.08 | 271.0814 | $C_{12}H_{15}O_7$ | 0.6 | 139.08 | arbutin | x | x | x | nd | x | x |
| 2 | 4.3 | 343.0658 | $C_{14}H_{15}O_{10}$ | −0.6 | 283.8/191.1 | galloyl quinic acid | x | x | x | x | x | x |
| 3 | 5.4 | 289.0708 | $C_{27}H_{21}O_{18}$ | −1.5 | 275.9/245.1/205.1/144.2 | catechin/epicatechin | x | x | x | x | x | x |
| 4 | 5.5 | 353.0857 | $C_{16}H_{17}O_9$ | −2.8 | 262.8/191.01/171.7 | chlorogenic acid | nd | nd | x | nd | nd | x |
| 5 | 5.8 | 633.0712 | $C_{27}H_{21}O_{18}$ | −1.5 | 463.06/300.9 | strictinin ellagitannin | nd | x | x | x | nd | x |
| 6 | 5.9 | 289.0706 | $C_{27}H_{21}O_{18}$ | −1.5 | 275.9/245.1/205.1/144.2 | catechin/epicatechin | x | x | x | x | x | x |
| 7 | 7.6 | 633.0709 | $C_{27}H_{21}O_{18}$ | −1.5 | 463.06/300.9 | strictinin ellagitannin | nd | x | x | x | nd | x |
| 8 | 8.0 | 477.0656 | $C_{21}H_{17}O_{13}$ | 0.2 | 462.0/366.07/325.05/191.03/164.7 | digalloylquinic shikimic acid | x | x | x | x | x | x |
| 9 | 10.9 | 479.0815 | $C_{21}H_{19}O_{13}$ | −1.43 | 451.2/354.7/316.02/297.6/165.8/145.5 | myricetin glucoside | nd | nd | x | nd | nd | x |
| 10 | 13.0 | 449.0705 | $C_{20}H_{17}O_{12}$ | −1.7 | 317.02/183.3/149.5 | myricetin pentoside | nd | x | x | nd | nd | x |
| 11 | 13.3 | 615.0967 | $C_{28}H_{23}O_{16}$ | 1.8 | 463.08/301.03 | quercetin galloylhexoside isomer | x | x | x | x | x | x |
| 12 | 14.2 | 463.0862 | $C_{21}H_{19}O_{12}$ | −1.5 | 316.02/284.8/183.5 | myricetin rhamnoside | nd | x | x | x | nd | x |
| 13 | 17.1 | 433.0758 | $C_{20}H_{17}O_{11}$ | −1.12 | 355.3/312.7/301.03/283.8/216.2 | quercetin pentoside | x | x | x | x | x | x |
| 14 | 17.2 | 615.0960 | $C_{28}H_{23}O_{16}$ | 1.02 | 463.08/301.03/265.3/241.08 | quercetin galloylhexoside isomer | x | x | x | x | x | x |
| 15 | 17.7 | 433.0758 | $C_{20}H_{17}O_{11}$ | −1.1 | 301.03/265.8/247.7/219.2/181.9/134.9 | quercetin pentoside | x | x | x | x | x | x |
| 16 | 19.5 | 463.0860 | $C_{21}H_{19}O_{12}$ | 0.8 | 404.3/316.02/301.04/282; 2 | isoquercitrin | nd | x | x | x | nd | x |
| 17 | 19.7 | 447.0913 | $C_{21}H_{19}O_{11}$ | −1.2 | 301.03/287.9/178.9 | quercitrin | x | x | x | x | x | x |
| 18 | 22.4 | 417.0810 | $C_{20}H_{17}O_{10}$ | −0.9 | 374.9/285.04/175.6 | kaempferol pentoside | nd | x | x | x | x | x |
| 19 | 24.7 | 431.0965 | $C_{21}H_{19}O_{10}$ | −1.09 | 285.03/235.2/195.9 | Kaempferol-rhamnoside (afzelin) | x | x | x | nd | x | x |

(L = Leaves; FR = Red Fruit; FG = Yellow Fruit; SS = Sassari; LM = La Maddalena). Compounds were putatively identified with ID level (Identification level): Putatively annotated compounds (e.g., without chemical reference standards, based upon physicochemical properties and/or spectral similarity with public/commercial spectral libraries).

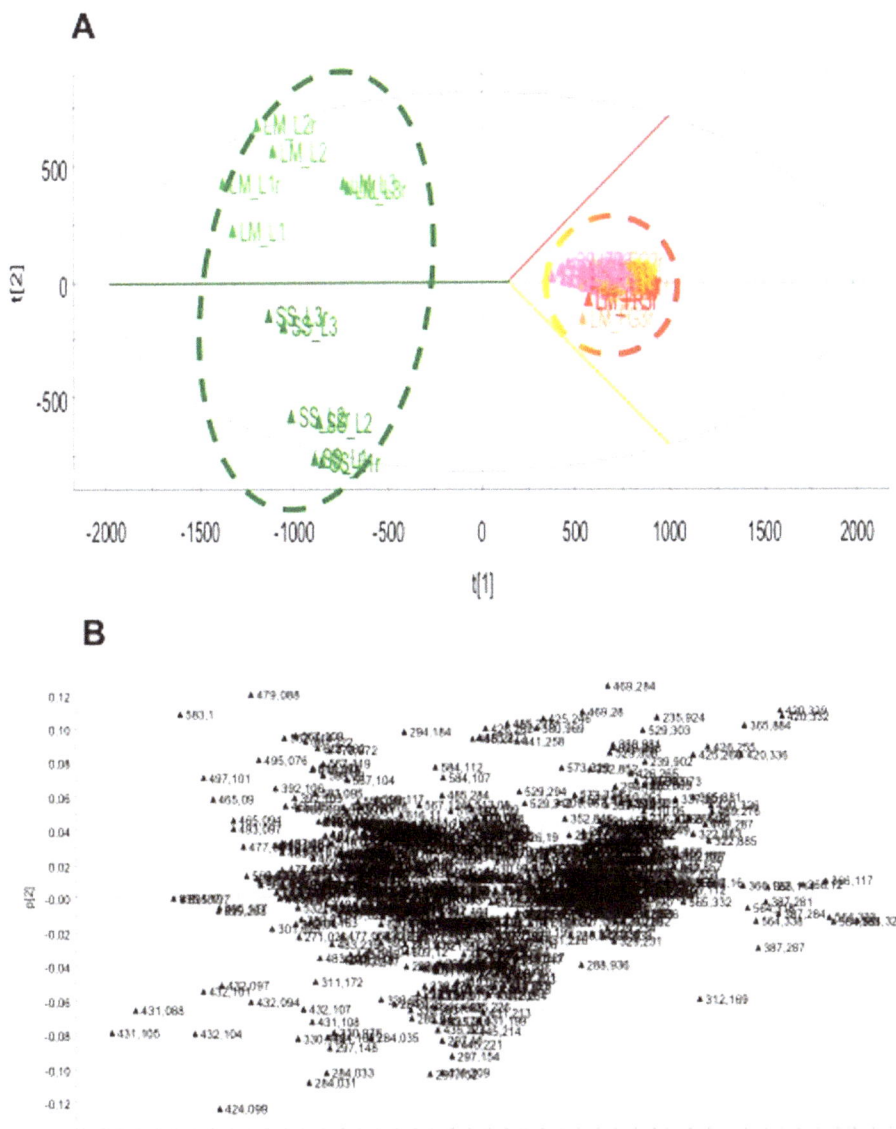

**Figure 4.** Score plot (**A**) and loading plot (**B**) obtained from HR–LCMS data. (L = Leaves; FR = Red Fruit; FG = Yellow Fruit; SS = Sassari; LM = La Maddalena).

## 3.3. Antioxidant Activity

Two different assays were employed to assess the antioxidant activities of different *A. unedo* extracts: (a) DPPH assay, a direct method based on the scavenging of the radical; and (b) ABTS assay, based on the inhibition of cation radical by antioxidants. IC50 value (the quantity of an extract able to neutralize the 50% of the radical) was taken into account to evaluate the antioxidant activity because the measurement of the absolute value for the antioxidant activity of an extract is contingent on many variables, comprising degradation during the analysis and matrix interference, and these contribute to the possibility of a wrong value. Table 2 reports the results obtained for both DPPH and ABTS assays. Regarding the DPPH assay, leaves extracts of both Sardinia's areas (Sassari and La Maddalena) showed an antioxidant activity significantly higher than red fruits and yellow fruits extracts, respectively, at time zero and after 30 min, a time when the reaction was stable and complete.

**Table 2.** Scavenging of 50% of DPPH and ABTS radical by Trolox and ethanolic extracts of *Arbutus unedo* of different areas of Sardinia, at different time points.

|  | DPPH IC$_{50}$ | | | ABTS IC$_{50}$ | | |
|---|---|---|---|---|---|---|
|  | 0 min | 30 min | *p* Value | 0 min | 50 min | *p* Value |
| Trolox (µg/mL) | 13.50 | 6.15 |  | 3.34 | 3.28 |  |
| *A. unedo* Sassari Red fruits (µg/mL) | 229.09 ± 81.75 | 103.59 ± 45.97 | Time 0 min LSS vs. FR SS *p* < 0.05 | 52.71 ± 16.13 | 21.43 ± 7.34 | Time 0 min LSS vs. FR SS *p* < 0.01 |
| *A. unedo* Sassari Yellow fruits (µg/mL) | 300.92 ± 107.31 | 131.53 ± 44.61 | LSS vs. FG SS *p* < 0.05 | 101.11 ± 53.44 | 40.33 ± 21.95 | LSS vs. FG SS *p* < 0.05 |
| *A. unedo* Sassari leaves (µg/mL) | 29.37 ± 5.87 | 15.18 ± 4.11 | Time 30 min LSS vs. FR SS *p* < 0.05 LSS vs. FG SS *p* < 0.01 | 7.63 ± 1.38 | 2.35 ± 0.06 | Time 50 min LSS vs. FR SS *p* < 0.01 LSS vs. FG SS *p* < 0.05 |
| *A. unedo* La Maddalena Red fruits (µg/mL) | 235.67 ± 43.72 | 116.84 ± 16.54 | Time 0 min L LM vs. FR LM *p* < 0.01 | 79.36 ± 46.09 | 34.91 ± 23.07 | Time 0 min L LM vs. FG LM *p* < 0.05 |
| *A. unedo* La Maddalena Yellow fruits (µg/mL) | 380.14 ± 90.72 | 180.30 ± 75.48 | L LM vs. FG LM *p* < 0.01 | 96.09 ± 40.98 | 49.89 ± 17.62 |  |
| *A. unedo* La Maddalena leaves (µg/mL) | 23.35 ± 4.25 | 8.85 ± 7.79 | Time 30 min L LM vs. FR LM *p* < 0.01 L LM vs. FG LM *p* < 0.05 | 6.73 ± 2.14 | 2.72 ± 1.68 | Time 50 min L LM vs. FG LM *p* < 0.01 |

Data were expressed as means ± SD. A *p* < 0.05 was considered significant. FR SS = Red fruits of *A. unedo* from Sassari; FG SS = Yellow fruits of *A. unedo* from Sassari; L SS = Leaves of *A. unedo* from Sassari; FR LM = Red fruits of *A. unedo* from La Maddalena; FG LM = Yellow fruits of *A. unedo* from La Maddalena; L LM = Leaves of *A. unedo* from La Maddalena.

La Maddalena leaves extracts presented a higher free radical scavenging effect than those from Sassari, even if not statistically significantly different. Similarity, the ABTS results were in agreement with those obtained for DPPH, even if ABTS gave the best results, showing for Sassari and La Maddalena leaves extracts IC50 values of 2.35 ± 0.06 and 2.72 ± 1.68 µg/mL, respectively, lower than those obtained for Trolox, reference compound.

The best results showed by ABTS assay over DPPH depend on the fact that ABTS can be used over a wider range of pH and is able to measure the antioxidant capacity of both water-soluble and lipid-soluble metabolites, since it can be dissolved in both aqueous and organic media, unlike DPPH, which can only be solubilized in alcoholic media [36].

Determination of total phenolic content was carried out for all extracts using Folin–Ciocalteu assay (Table 3). Obtained results were in agreement with the demonstrated antioxidant activity of the three ethanolic extracts obtained from samples from different areas of Sardinia.

Table 3. Determination of phenols by Folin–Ciocalteu's method.

| Extract Concent Ration (µg/mL) | GAE A. Unedo Sassari Red Fruits (µg GAE) | GAE A. Unedo Sassari Yellow Fruits (µg GAE) | GAE A. Unedo Sassari Leaves (µg GAE) | GAE A. Unedo La Maddalena Red Fruits (µg GAE) | GAE A. Unedo La Maddalena Yellow Fruits (µg GAE) | GAE A. Unedo La Maddalena Leaves(µg GAE) | Pearson Product Moment Correlation |
|---|---|---|---|---|---|---|---|
| 100 | 67.10 ± 17.14 *† | 43.62 ± 11.85 *† | 347.56 ± 126.05 | 53.16 ± 35.75 *† | 50.62 ± 20.64 *† | 327.90 ± 126.75 | |
| 50 | 35.33 ± 9.37 *† | 22.72 ± 6.41 *† | 179.52 ± 70.68 | 26.42 ± 11.73 *† | 26.42 ± 8.04 *† | 181.55 ± 84.65 | * vs. DPPH |
| 25 | 23.67 ± 4.29 *† | 16.13 ± 9.12 *† | 116.46 ± 36.12 | 21.11 ± 11.83 *† | 18.10 ± 5.64 *† | 143.84 ± 67.03 | $p < 0.01$ |
| 10 | 10.15 ± 3.27 *† | 6.78 ± 2.43 *† | 51.50 ± 15.93 | 9.53 ± 3.44 *† | 8.30 ± 2.42 *† | 64.56 ± 33.28 | † vs. ABTS |
| 5 | 6.63 ± 1.16 *† | 4.24 ± 0.55 *† | 29.01 ± 4.24 | 6.17 ± 3.36 *† | 5.04 ± 3.89 *† | 38.12 ± 17.86 | $p < 0.05$ |
| 1 | 3.17 ± 1.66 *† | 2.40 ± 0.70 *† | 10.04 ± 1.54 | 4.91 ± 1.69 *† | 2.52 ± 1.17 *† | 18.54 ± 14.44 | |

* $p < 0.01$, † $p < 0.05$.

Data were expressed as means ± SD of three independent experiments. Each result of red and yellow fruits showed a positive correlation with DPPH (* $p < 0.01$) and ABTS results († $p < 0.05$).

However, a positive Pearson correlation between antioxidant activity measured with DPPH and ABTS and total phenolic contents was detected only in red and yellow fruits, probably because in leaves, antioxidant activity measured with DPPH and ABTS reached saturation early as low concentrations, due to the occurrence of different compounds in leaves extracts. Regardless, the higher amount of total phenols of leaves as well as red fruits and yellow fruits demonstrated that antioxidant activity is governed by phenolic amounts, confirming previous data [37].

## 4. Conclusions

In conclusion, in a totally untargeted approach, the power of HPTLC in discriminating origin and organs of samples was evaluated. The obtained results were compared with a more performant as well as more complex technique, mainly HRMS. Comparing the results obtained for both the techniques and visualized by PCA, the discrimination results are comparable. Then, for a preliminary untargeted approach, by integrating HPTLC with PCA, it's possible to discriminate samples with good confidence and in a short time.

The second aim of the paper was to carry out a qualitative analysis of the main compounds occurring in the samples from different origins and organs. This objective was reached by data collected by LC–HRMS, which permitted it to assign an elemental formula for the main $m/z$ value and LC–HR–MS (MS) to enrich information with a fragmentation pattern useful to complete identification of compounds.

Finally, we also confirm that the high antioxidant activity is related to the amount of total phenolic content.

The present study highlighted that the qualitative and quantitative chemical diversity detected by analytical techniques, related to the slightly different habitats in which *Arbutus unedo* grow, give rise, after multivariate analysis, to a discrimination of the studied samples. HPTLC and LC–ESI–Orbitrap–MS provide different and complementary data and together these may be employed to really differentiate between a wide variety of crude drug powders and herbal medicinal products.

**Author Contributions:** Conceptualization, M.M. and P.M.; Methodology, G.D.U. and G.P.; Software, M.M., G.D.U., P.M., and G.P.; Validation, D.C. and M.M.; Formal analysis, G.D.U., M.M., and G.P.; Investigation, G.D.U., M.M., and G.P.; Resources, G.L.P., M.F., and M.M.; Writing—original draft preparation, M.M., G.D.U., G.L.P., and F.R.G.; Writing—review and editing, D.C., G.P., G.M., and P.M.; Supervision, M.M. and P.M.; Project administration, M.M. and G.P.

**Funding:** This research did not receive any specific grant from funding agencies in the public, commercial, or not-for-profit sectors.

**Conflicts of Interest:** The authors declare no conflict of interest.

## References

1. Astarita, G.; Langridge, J. An Emerging Role for Metabolomics in Nutrition Science. *J. Nutr. Nutr.* **2013**, *6*, 181–200. [CrossRef] [PubMed]
2. Chen, W.; Gong, L.; Guo, Z.L.; Wang, W.S.; Zhang, H.Y.; Liu, X.Q.; Yu, S.B.; Xiong, L.Z.; Luo, J. A Novel Integrated Method for Large-Scale Detection, Identification, and Quantification of Widely Targeted Metabolites: Application in the Study of Rice Metabolomics. *Mol. Plant* **2013**, *6*, 1769–1780. [CrossRef] [PubMed]
3. Duan, L.X.; Chen, T.L.; Li, M.; Chen, M.; Zhou, Y.Q.; Cui, G.H.; Zhao, A.H.; Jia, W.; Huang, L.Q.; Qi, X.Q. Use of the Metabolomics Approach to Characterize Chinese Medicinal Material Huangqi. *Mol. Plant* **2012**, *5*, 376–386. [CrossRef] [PubMed]
4. Kueger, S.; Steinhauser, D.; Willmitzer, L.; Giavalisco, P. High-resolution plant metabolomics: From mass spectral features to metabolites and from whole-cell analysis to subcellular metabolite distributions. *Plant J.* **2012**, *70*, 39–50. [CrossRef] [PubMed]
5. Booker, A.; Frommenwiler, D.; Johnston, D.; Umealajekwu, C.; Reich, E.; Heinrich, M. Chemical variability along the value chains of turmeric (*Curcuma longa*): A comparison of nuclear magnetic resonance spectroscopy and high performance thin layer chromatography. *J. Ethnopharmacol.* **2014**, *152*, 292–301. [CrossRef]
6. Holtorf, H.; Guitton, M.C.; Reski, R. Plant functional genomics. *Naturwissenschaften* **2002**, *89*, 235–249. [CrossRef] [PubMed]
7. Sumner, L.W.; Mendes, P.; Dixon, R.A. Plant metabolomics: Large-scale phytochemistry in the functional genomics era. *Phytochemistry* **2003**, *62*, 817–836. [CrossRef]
8. Tolstikov, V.V.; Fiehn, O. Analysis of highly polar compounds of plant origin: Combination of hydrophilic interaction chromatography and electrospray ion trap mass spectrometry. *Anal. Biochem.* **2002**, *301*, 298–307. [CrossRef]
9. Vorst, O.; De Vos, C.H.R.; Lommen, A.; Staps, R.V.; Visser, R.G.F.; Bino, R.J.; Hall, R.D. A non-directed approach to the differential analysis of multiple LC-MS-derived metabolic profiles. *Metabolomics* **2005**, *1*, 169–180. [CrossRef]
10. Avula, B.; Sagi, S.; Gafner, S.; Upton, R.; Wang, Y.H.; Wang, M.; Khan, I.A. Identification of Ginkgo biloba supplements adulteration using high performance thin layer chromatography and ultra high performance liquid chromatography-diode array detector-quadrupole time of flight-mass spectrometry. *Anal. Bioanal. Chem.* **2015**, *407*, 7733–7746. [CrossRef]
11. Piccinonna, S.; Ragone, R.; Stocchero, M.; Del Coco, L.; De Pascali, S.A.; Schena, F.P.; Fanizzi, F.P. Robustness of NMR-based metabolomics to generate comparable data sets for olive oil cultivar classification. An inter-laboratory study on Apulian olive oils. *Food Chem.* **2016**, *199*, 675–683. [CrossRef]
12. Zhang, L.; Zeng, Z.D.; Zhao, C.X.; Kong, H.W.; Lu, X.; Xu, G.W. A comparative study of volatile components in green, oolong and black teas by using comprehensive two-dimensional gas chromatography-time-of-flight mass spectrometry and multivariate data analysis. *J Chromatogr. A* **2013**, *1313*, 245–252. [CrossRef]
13. Zhao, Y.; Chen, P.; Lin, L.Z.; Harnly, J.M.; Yu, L.L.; Li, Z.W. Tentative identification, quantitation, and principal component analysis of green pu-erh, green, and white teas using UPLC/DAD/MS. *Food Chem.* **2011**, *126*, 1269–1277. [CrossRef]
14. Diniz, P.H.G.D.; Barbosa, M.F.; Milanez, K.D.T.D.; Pistonesi, M.F.; De Araujo, M.C.U. Using UV-Vis spectroscopy for simultaneous geographical and varietal classification of tea infusions simulating a home-made tea cup. *Food Chem.* **2016**, *192*, 374–379. [CrossRef]
15. Scoparo, C.T.; De Souza, L.M.; Dartora, N.; Sassaki, G.L.; Gorin, P.A.J.; Iacomini, M. Analysis of Camellia sinensis green and black teas via ultra high performance liquid chromatography assisted by liquid-liquid partition and two-dimensional liquid chromatography (size exclusion x reversed phase). *J Chromatogr. A* **2012**, *1222*, 29–37. [CrossRef]
16. Maldini, M.; Montoro, P.; Addis, R.; Toniolo, C.; Petretto, G.L.; Foddai, M.; Nicoletti, M.; Pintore, G. A new approach to discriminate *Rosmarinus officinalis* L. plants with antioxidant activity, based on HPTLC fingerprint and targeted phenolic analysis combined with PCA. *Ind. Crop. Prod.* **2016**, *94*, 665–672. [CrossRef]
17. Martelanc, M.; Vovk, I.; Simonovska, B. Separation and identification of some common isomeric plant triterpenoids by thin-layer chromatography and high-performance liquid chromatography. *J. Chromatogr. A* **2009**, *1216*, 6662–6670. [CrossRef]

18. Lakavath, S.; Avula, B.; Wang, Y.H.; Rumalla, C.S.; Gandhe, S.; Venkatrao, A.R.B.; Satishchandra, P.A.; Bobbala, R.K.; Khan, I.A.; Narasimha, A.R.A.V. Differentiating the Gum Resins of Two Closely Related Indian Gardenia Species, *G. gummifera* and *G-lucida*, and Establishing the Source of Dikamali Gum Resin Using High-Performance Thin-Layer Chromatography and Ultra-Performance Liquid Chromatography-UV/MS. *J. AOAC Int.* **2012**, *95*, 67–73. [CrossRef]
19. Reich, E.; Schibli, A. *High-Performance Thin Layer Chromatography for the Analysis of Medicinal Plants*; Thieme: New York, NY, USA, 2007.
20. Ogegbo, O.L.; Eyob, S.; Parmar, S.; Wang, Z.T.; Bligh, S.W.A. Metabolomics of four TCM herbal products: Application of HPTLC analysis. *Anal. Methods UK* **2012**, *4*, 2522–2527. [CrossRef]
21. Audoin, C.; Holderith, S.; Romari, K.; Thomas, O.P.; Genta-Jouve, G. Development of a Work-Flow for High-Performance Thin-Layer Chromatography Data Processing for Untargeted Metabolomics. *JPC J. Planar Chromatogr.* **2014**, *27*, 328–332. [CrossRef]
22. Wang, J.; Cao, X.S.; Qi, Y.D.; Ferchaud, V.; Chin, K.L.; Tang, F. High-Performance Thin-Layer Chromatographic Method for Screening Antioxidant Compounds and Discrimination of Hibiscus sabdariffa L. by Principal Component Analysis. *JPC J. Planar Chromatogr.* **2015**, *28*, 274–279. [CrossRef]
23. Tuberoso, C.I.G.; Bifulco, E.; Caboni, P.; Cottiglia, F.; Cabras, P.; Floris, I. Floral Markers of Strawberry Tree (*Arbutus unedo* L.) Honey. *J. Agric. Food Chem.* **2010**, *58*, 384–389. [CrossRef]
24. Kontogianni, V.G.; Tomic, G.; Nikolic, I.; Nerantzaki, A.A.; Sayyad, N.; Stosic-Grujicic, S.; Stojanovic, I.; Gerothanassis, I.P.; Tzakos, A.G. Phytochemical profile of Rosmarinus officinalis and Salvia officinalis extracts and correlation to their antioxidant and anti-proliferative activity. *Food Chem.* **2013**, *136*, 120–129. [CrossRef]
25. Van De Velde, F.; Tarola, A.M.; Guemes, D.; Pirovani, M.E. Bioactive compounds and antioxidant capacity of Camarosa and Selva strawberries (*Fragaria x ananassa* Duch.). *Foods* **2013**, *2*, 120–131. [CrossRef]
26. Pluskal, T.; Castillo, S.; Villar-Briones, A.; Oresic, M. MZmine 2: Modular framework for processing, visualizing, and analyzing mass spectrometry-based molecular profile data. *BMC Bioinform.* **2010**, *11*, 395. [CrossRef]
27. D'urso, G.; Pizza, C.; Piacente, S.; Montoro, P. Combination of LC-MS based metabolomics and antioxidant activity for evaluation of bioactive compounds in Fragaria vesca leaves from Italy. *J. Pharm. Biomed.* **2018**, *150*, 233–240. [CrossRef]
28. Brand-Williams, W.; Cuvelier, M.E.; Berset, C. Use of a Free-Radical Method to Evaluate Antioxidant Activity. *Food Sci. Technol. Leb* **1995**, *28*, 25–30. [CrossRef]
29. Petretto, G.L.; Maldini, M.; Addis, R.; Chessa, M.; Foddai, M.; Rourke, J.P.; Pintore, G. Variability of chemical composition and antioxidant activity of essential oils between Myrtus communis var. Leucocarpa DC and var. Melanocarpa DC. *Food Chem.* **2016**, *197*, 124–131. [CrossRef]
30. Lizcano, L.J.; Bakkali, F.; Ruiz-Larrea, M.B.; Ruiz-Sanz, J.I. Antioxidant activity and polyphenol content of aqueous extracts from Colombian Amazonian plants with medicinal use. *Food Chem.* **2010**, *119*, 1566–1570. [CrossRef]
31. Jardim, C.; Macedo, D.; Figueira, I.; Dobson, G.; Mcdongall, G.J.; Stewart, D.; Ferreira, R.B.; Menezes, R.; Santos, C.N. (Poly)phenol metabolites from *Arbutus unedo* leaves protect yeast from oxidative injury by activation of antioxidant and protein clearance pathways. *J. Funct. Foods* **2017**, *32*, 333–346. [CrossRef]
32. Pallauf, K.; Rivas-Gonzalo, J.C.; Del Castillo, M.D.; Cano, M.P.; De Pascual-Teresa, S. Characterization of the antioxidant composition of strawberry tree (*Arbutus unedo* L.) fruits. *J. Food Compos. Anal.* **2008**, *21*, 273–281. [CrossRef]
33. Tuberoso, C.I.G.; Boban, M.; Bifulco, E.; Budimir, D.; Pirisi, F.M. Antioxidant capacity and vasodilatory properties of Mediterranean food: The case of Cannonau wine, myrtle berries liqueur and strawberry-tree honey. *Food Chem.* **2013**, *140*, 686–691. [CrossRef]
34. Nenadis, N.; Llorens, L.; Koufogianni, A.; Diaz, L.; Font, J.; Gonzalez, J.A.; Verdaguer, D. Interactive effects of UV radiation and reduced precipitation on the seasonal leaf phenolic content/composition and the antioxidant activity of naturally growing *Arbutus unedo* plants. *J. Photochem. Photobiol. B* **2015**, *153*, 435–444. [CrossRef]
35. Fortalezas, S.; Tavares, L.; Pimpao, R.; Tyagi, M.; Pontes, V.; Alves, P.M.; Mcdougall, G.; Stewart, D.; Ferreira, R.B.; Santos, C.N. Antioxidant Properties and Neuroprotective Capacity of Strawberry Tree Fruit (*Arbutus unedo*). *Nutrients* **2010**, *2*, 214–229. [CrossRef]

36. Longhi, J.G.; Perez, E.; De Lima, J.J.; Candido, L.M.B. In vitro evaluation of *Mucuna pruriens* (L.) DC. antioxidant activity. *Braz. J. Pharm. Sci.* **2011**, *47*, 535–544. [CrossRef]
37. Erkan, N.; Ayranci, G.; Ayranci, E. Antioxidant activities of rosemary (*Rosmarinus Officinalis*, L.) extract, blackseed (*Nigella sativa* L.) essential oil, carnosic acid, rosmarinic acid and sesamol. *Food Chem.* **2008**, *110*, 76–82. [CrossRef]

© 2019 by the authors. Licensee MDPI, Basel, Switzerland. This article is an open access article distributed under the terms and conditions of the Creative Commons Attribution (CC BY) license (http://creativecommons.org/licenses/by/4.0/).

Article

# Two-Way Characterization of Beekeepers' Honey According to Botanical Origin on the Basis of Mineral Content Analysis Using ICP-OES Implemented with Multiple Chemometric Tools

Artemis Panormitis Louppis [1], Ioannis Konstantinos Karabagias [2,*], Chara Papastephanou [1] and Anastasia Badeka [2]

[1] CP. Foodlab Ltd., Polifonti 25, Strovolos, 2047 Nicosia, Cyprus; artemislouppis@gmail.com (A.P.L.); foodlab@cytanet.com.cy (C.P.)
[2] Laboratory of Food Chemistry, Department of Chemistry University of Ioannina, 45110 Ioannina, Greece; abadeka@uoi.gr
* Correspondence: ioanniskarabagias@gmail.com or ikaraba@cc.uoi.gr; Tel.: +30-697-828-6866

Received: 27 May 2019; Accepted: 12 June 2019; Published: 14 June 2019

**Abstract:** Asfaka, fir, flower, forest flowers and orange blossom honeys harvested in the wider area of Hellas by professional beekeepers, were subjected to mineral content analysis using inductively coupled plasma optical emission spectrometry (ICP-OES). The main purpose of this study was to characterize the mineral profile and content of toxic metals such as lead, cadmium and chromium, and investigate whether specific minerals could assist accurately in the botanical origin discrimination with implementation of chemometrics. Twenty-five minerals were identified (Ag, Al, As, B, Ba, Be, Ca, Cd, Co, Cr, Cu, Fe, Hg, Mg, Mn, Mo, Ni, Pb, Sb, Se, Si, Ti, Tl, V, Zn) and quantified. Results showed that the mineral content varied significantly ($p < 0.05$) according to honey botanical origin, whereas lead, cadmium, and chromium contents ranged between 0.05–0.33 mg kg$^{-1}$, <0.05 mg kg$^{-1}$, and in the range of <0.12 to 0.39 mg kg$^{-1}$, respectively. Fir honeys from Aitoloakarnania region showed the highest mineral content (182.13 ± 71.34 mg kg$^{-1}$), while flower honeys from Samos Island recorded the highest silicon content (16.08 ± 2.94 mg kg$^{-1}$). Implementation of multivariate analysis of variance (MANOVA), factor analysis (FA), linear discriminant analysis (LDA), and stepwise discriminant analysis (SDA) led to the perfect classification (100%) of these honeys according to botanical origin with the use of Al, As, Ca, Mg, Mn, Ni, Pb, Sb, Si, Zn and total mineral content. However, the higher lead content in the majority of samples than the regulated upper limit (0.10 mg kg$^{-1}$), sets the need for further improvements of the beekeepers' practices/strategies for honey production.

**Keywords:** characterization; beekeepers' honey; minerals; inductively coupled plasma optical emission spectrometry (ICP-OES); chemometrics

## 1. Introduction

*Apis mellifera* honeybees produce honey, a delicious natural sweetener, from nectar or honeydew secretions by adding specific substances of their own (i.e., enzymes). The basic honey components comprise sugars and water, whereas other minor components such as minerals, organic acids, enzymes, amino acids, polyphenols, fatty acids, pollen and wax originate by either bees or plants, and show diverse amounts in relation to honey botanical and geographical origin [1].

Honey may be classified into two main categories: blossom and honeydew. Typical examples of blossom honeys are thyme, citrus and heather honeys, while honeydew honeys are pine, fir, and oak honeys. However, there are numerous honey types familiar to the research community, but less known to consumers' society. These include, for example, prairie and native vegetation honeys [2], coffee honey [3], rape, dandelion and rhododendron honeys [4], and forest and clover honeys [5].

The botanical origin of honey is officially confirmed by using the microscopic data of its pollen content (melissopalynology) [6]. In addition, the European regulation 2001/110/EC [7] sets the basic criteria for honey quality and identity. For instance, it is strictly stated that both the botanical and geographical origin of honey must be declared on the package label. Furthermore, this directive also sets specific compositional characteristics such as moisture, hydroxymethyl furfural content (HMF), free acid content, electrical conductivity, and enzymatic activity, etc. that should be taken into account.

In the literature, however, there have been numerous studies focusing on the determination of physicochemical parameters [8,9], volatile compounds [10], polyphenols [11,12], minerals [2,13–16], free amino acids [17], organic acids [17], etc., for the recognition of honey geographical and botanical origin using either official or instrumental techniques in combination with chemometrics.

Honey has a low mineral content (ca. 0.1–0.2% in blossom honeys and 1% or higher in honeydew honeys); however, mineral content analysis has gained considerable attention. Potassium has been reported to be the dominant mineral in honey, followed by other minerals such as sodium, phosphorus, magnesium, manganese, iron, copper, silicon and other trace elements [2].

There are different techniques that have been applied for the determination of honey mineral content. The most commonly used are inductively coupled plasma sector field mass spectrometry [4], inductively coupled plasma mass spectroscopy [14], inductively coupled plasma optical emission spectrometry (ICP-OES) [5,15,16], flame photometry and atomic absorption spectroscopy [2].

More recently, the European Commission Regulation [18], in an effort to guarantee honey quality and consumer safety, set the maximum level of lead to be 0.1 mg kg$^{-1}$. However, numerous honey types are sold in the Hellenic and global market and are not mentioned, for example, in the packaging labels in regards to the levels of certain toxic metals such as lead or cadmium. In addition, there is also lack of melissopalynological analysis results. For these honeys it is usually declared in the packaging label that only the botanical and geographical origin of honey is in conformity; hence, adherent to the EU Council directive [7].

However, due to the press of the global market with products of low quality and of no specific origin, physico-chemical "marker" analysis in combination with chemometrics may effectively assist in the determination of honey botanical and geographical origin and contribute to the establishment of health safety standards [8,9,15,19].

Based on the aforementioned, the objective of the present study was to investigate whether mineral content analysis using ICP-OES could effectively assist in the botanical origin differentiation of some lesser known Hellenic honey types such as asfaka, cotton, fir, flower, forest flowers and orange blossom honeys obtained directly from beekeepers, and provide information about lead and cadmium content of these honeys in terms of health safety standards, as relevant data for these honey types had not been reported previously.

## 2. Materials and Methods

### 2.1. Honey Samples

Due to the extent of natural limitations, nineteen honey samples ($n = 19$) of different botanical and geographical origins were collected directly from beekeepers. Flower ($n = 3$), fir ($n = 3$) and asfaka ($n = 2$) honeys were collected from Aitoloakarnania. Additional flower honey samples were collected from Samos Island ($n = 7$). The orange blossom honeys ($n = 2$) were collected from Lakonia and forest flowers honeys ($n = 2$) were collected from Zagorochoria (Ioannina, Hellas, Greece). The samples were stored in glass containers and maintained at 4 ± 1 °C until analysis.

## 2.2. Reagents and Solutions

The chemicals used in the present study were of analytical reagent grade. All standard solutions were prepared by appropriate dilution of a multi-element standard (100 mg L$^{-1}$) obtained from Merck (Darmstadt, Germany). The Hg standard solution (1000 mg L$^{-1}$) was purchased from Sigma-Aldrich (Darmstadt, Germany). Nitric acid (suprapure 65%) was obtained from Merck (Darmstadt, Germany). All other solutions and dilutions were prepared with ultrapure water (Milli-Q, Millipore, Bedford, MA, USA) [15].

## 2.3. Preparation of Honey Samples for ICP-OES Analysis

Approximately 10 g of each honey sample were digested and mineralized by adding 65% concentrated HNO$_3$ (10 mL) and by heating the mixture in a water bath at 60 °C for 30 min. The mixture was then sonicated and diluted to a final volume of 100 mL with ultrapure water before ICP-OES analysis [15]. Each analysis was carried out in triplicate ($n$ = 3) and results reported are the average ± standard deviation values.

## 2.4. ICP-OES Instrumentation and Conditions

Mineral analysis was carried out using a Thermo Scientific IRIS Intrepid II XDL inductively coupled plasma-atomic emission spectrometer (Thermo Electron Corporation, Waltham, MA, USA). The emission wavelength for each element, multi-elemental analysis parameters, and method analytical characteristics are given in previous studies [15,16].

## 2.5. Chemometric Techniques

All data processing and implementation of the chemometric techniques followed were performed using the Statistical Package for the Social Sciences (SPSS) version 20.0 (IBM Corp. Armonk, NY, USA).

### 2.5.1. Multivariate Analysis of Variance (MANOVA)

MANOVA may be characterized as a generalized form of univariate analysis of variance (ANOVA); although, unlike univariate ANOVA, MANOVA uses the covariance between outcome variables in testing the statistical significance of the mean differences [20]. It then implies a pre-evaluation step of the significance of the measured variables since it defines only the significant variables ($p < 0.05$) (considering the possible effectiveness of all variables simultaneously) assisting in the classification of samples [21].

### 2.5.2. Factor Analysis (FA)

Factor analysis (FA) is a multivariate statistical technique used to describe the variability among correlated variables in terms of a potentially lower number of unobserved variables called factors. Factor analysis searches for such joint variations in response to unobserved, independent, and latent variables. The observed variables are modelled as linear combinations of the potential factors, plus the *error* terms. Both PCA and FA aim to reduce the dimensionality of a set of data, but the approaches followed are quite different for the two techniques. Factor analysis is clearly designed with the objective to identify certain unobservable factors from the observed variables, whereas PCA does not directly address this objective; PCA provides an approximation to the required factors [22]. There are numerous procedures designed to determine the optimal number of factors to retain in FA. In the present work, the Kaiser's eigenvalue-greater-than-one rule (or K1 rule) was considered.

### 2.5.3. Linear Discriminant Analysis (LDA)

LDA is a pattern recognition technique and is usually implemented after the application of MANOVA analysis or after the use of non-supervised techniques such as PCA, FA, etc. LDA uses the significant parameters defined by MANOVA (independent variables) ($p < 0.05$) to determine a linear combination of these groups of subjects which could differentiate the "a priori" known groups (grouping variables) providing exact classification rates based on the use of original and cross validation methods [16].

2.5.4. Stepwise Discriminant Analysis (SDA)

In cases where a lot of predictors have arisen during the multivariate analysis, the stepwise method can be useful by automatically selecting the best variables to use in the model. We may term this procedure as an effective *haircut* of the discrimination variables for the predicting model efficacy. The stepwise method starts with a model that doesn't include any of the predictors. At each step, the predictor with the largest F to Enter value that exceeds the entry criteria (by default, 3.84 in the SPSS program) is added to the model. The variables left out of the analysis at the last step all have F to Enter values smaller than 3.84, so no more are added. The F to Remove values are useful for describing what really happens if a variable is removed from the current model (given that the other variables remain). F to Remove value for the entering variable is the same as that of F to Enter at the previous step [23].

## 3. Results

*3.1. Mineral Content of Asfaka, Cotton, Fir, Flower, Forest Flowers and Orange Blossom Honeys*

3.1.1. Abundant Minerals

The mineral content (mg kg$^{-1}$) of the analyzed honey samples varied significantly ($p < 0.05$) according to botanical origin (Table 1). Total mineral content followed this order: Fir honey from Aitoloakarnania > forest flowers honey from Zagorochoria > cotton honey from Larissa > flower honey from Samos Island > flower honey from Aitoloakarnania > asfaka honey from Aitoloakarnania. It has been documented in the literature that honeys containing honeydew secretions or possess dark colour (i.e. honeydew, fir, forest flowers, etc.) contain a higher content of minerals compared to nectar honeys (i.e. flower, asfaka, orange blossom, thyme, etc.) [4,5,13–16].

In all cases, the most abundant minerals were Ca, Mg, Si, and B, followed by Mn, Fe, and Zn.

Forest flowers honeys from Zagorochoria recorded the higher Ca content (58.30 ± 2.24, mg kg$^{-1}$), whereas the highest Mg content (81.71 ± 1.43, mg kg$^{-1}$), was recorded in fir honey samples from Aitoloakarnania, and may be indicative of these honey types grown in specific regions. On the contrary, Si recorded the highest content (16.08 ± 2.94, mg kg$^{-1}$) in flower honeys from Samos Island. This is also a finding that leads to the impact of both the botanical and geographical origins of honey mineral content, given the fact that the flower honeys from Aitoloakarnania, or even the other honey types studied, did not show this trend. This observation is in agreement with the results reported by Bogdanov et al. [4].

Lesser amounts of Cu, Al, As, Ni, Pb, and Sb were recorded in all honey samples. In addition, the content of Pb, As, Cu, and Ni was below 1 mg kg$^{-1}$ (Table 1). Only one flower honey sample from Samos Island (No.1) exceeded the value of 1 mg kg$^{-1}$ in Cu content.

The European Commission Regulation [18] has recently set the maximum level of 0.1 mg kg$^{-1}$ for lead in honey in an effort to ensure consumers safety and product's quality. Lead or Cd is flourishing in the natural environment due to human activity, and honey is often subjected to contamination. Regarding Pb content, the latter exceeded this new upper limit in flower honey samples from Aitoloakarnania and Samos Island, along with those of fir honeys from Aitoloakarnania.

The content values (mg kg$^{-1}$) obtained for Mg and Cu are within the range reported for different honey cultivars (rosemary, heather, orange blossom, and eucalyptus) cultivated in the wider area of Spain [13]. In addition, Mg and Cu content values of the present study are in agreement with the average ± standard deviation values reported for coffee honey from Brazil (82.01 ± 2.25 and 0.47 ± 0.04 mg kg$^{-1}$, respectively).

**Table 1.** Mineral content (mg kg$^{-1}$) of asfaka, cotton, fir, flower, forest flowers, and orange blossom honeys.

| Geographical Origin | Botanical Origin | Al | As | B | Ca | Cu | Fe | Mg | Mn | Ni | Pb | Sb | Si | Zn | Total Minerals (mg kg$^{-1}$) |
|---|---|---|---|---|---|---|---|---|---|---|---|---|---|---|---|
| Aitoloakarnania | Flower | 0.50 | 0.63 | 4.24 | 22.00 | 0.19 | 1.14 | 11.72 | 0.39 | 0.00 | 0.26 | 0.58 | 0.89 | 0.48 | 43.03 |
| Aitoloakarnania | Flower | 0.89 | 0.65 | 5.91 | 42.65 | 0.27 | 1.38 | 13.03 | 0.28 | 0.00 | 0.26 | 0.58 | 0.96 | 0.86 | 67.73 |
| Aitoloakarnania | Flower | 25.01 | 0.63 | 3.10 | 18.31 | 1.05 | 4.46 | 88.68 | 5.36 | 0.48 | 0.21 | 0.58 | 0.89 | 1.65 | 150.41 |
| | Average | 8.80 | 0.64 | 4.42 | 27.65 | 0.51 | 2.33 | 37.81 | 2.01 | 0.16 | 0.24 | 0.58 | 0.91 | 1.00 | 87.06 |
| | ±SD | 14.04 | 0.01 | 1.41 | 13.12 | 0.47 | 1.85 | 44.06 | 2.90 | 0.28 | 0.03 | 0.00 | 0.04 | 0.59 | 56.24 |
| Aitoloakarnania | Asfaka | 1.79 | 0.08 | 3.43 | 11.74 | 0.35 | 0.95 | 14.43 | 0.34 | 0.06 | 0.06 | 0.16 | 4.29 | 0.65 | 38.33 |
| Aitoloakarnania | Asfaka | 2.97 | 0.07 | 3.03 | 13.94 | 0.34 | 1.44 | 14.49 | 0.40 | 0.36 | 0.04 | 0.14 | 4.74 | 0.71 | 42.68 |
| | Average | 2.38 | 0.08 | 3.23 | 12.84 | 0.35 | 1.19 | 14.46 | 0.37 | 0.21 | 0.05 | 0.15 | 4.52 | 0.68 | 40.50 |
| | ±SD | 0.84 | 0.01 | 0.28 | 1.56 | 0.01 | 0.34 | 0.05 | 0.04 | 0.21 | 0.01 | 0.01 | 0.32 | 0.04 | 3.07 |
| Samos Island | Flower | 2.75 | 0.70 | 3.28 | 40.92 | 1.12 | 7.68 | 62.19 | 0.55 | 0.32 | 0.47 | 0.52 | 15.75 | 1.02 | 137.26 |
| Samos Island | Flower | 1.97 | 0.53 | 2.16 | 25.63 | 0.58 | 3.15 | 44.82 | 0.30 | 0.19 | 0.41 | 0.28 | 13.90 | 0.85 | 94.78 |
| Samos Island | Flower | 1.94 | 0.67 | 3.02 | 34.34 | 0.67 | 3.00 | 54.81 | 0.47 | 0.25 | 0.17 | 0.44 | 18.85 | 0.59 | 119.21 |
| Samos Island | Flower | 2.60 | 0.82 | 2.90 | 41.16 | 0.88 | 3.48 | 66.93 | 0.91 | 0.27 | 0.24 | 0.83 | 19.01 | 0.42 | 140.44 |
| Samos Island | Flower | 2.61 | 0.55 | 2.33 | 24.46 | 0.44 | 4.31 | 44.59 | 0.56 | 0.18 | 0.43 | 0.36 | 12.79 | 0.66 | 94.27 |
| Samos Island | Flower | 1.53 | 0.55 | 2.20 | 23.91 | 0.55 | 2.49 | 41.69 | 0.41 | 0.19 | 0.23 | 0.48 | 12.95 | 0.46 | 87.64 |
| Samos Island | Flower | 3.07 | 0.82 | 3.11 | 35.90 | 0.76 | 6.22 | 63.31 | 0.89 | 0.27 | 0.39 | 0.58 | 19.28 | 0.74 | 135.33 |
| | Average | 2.35 | 0.66 | 2.71 | 32.33 | 0.71 | 4.33 | 54.05 | 0.58 | 0.24 | 0.33 | 0.50 | 16.08 | 0.68 | 115.56 |
| | ±SD | 0.55 | 0.13 | 0.47 | 7.60 | 0.23 | 1.92 | 10.38 | 0.23 | 0.05 | 0.12 | 0.18 | 2.94 | 0.21 | 22.94 |
| Lakonia | Orange blossom | 0.40 | 0.07 | 4.26 | 30.19 | 0.11 | 0.59 | 19.30 | 0.40 | 0.05 | 0.08 | 0.16 | 4.69 | 2.21 | 62.51 |
| Lakonia | Orange blossom | 1.33 | 0.06 | 3.83 | 33.85 | 0.10 | 4.76 | 18.40 | 0.47 | 0.20 | 0.06 | 0.13 | 4.50 | 2.21 | 69.88 |
| | Average | 0.87 | 0.07 | 4.04 | 32.02 | 0.11 | 2.68 | 18.85 | 0.43 | 0.13 | 0.07 | 0.14 | 4.59 | 2.21 | 66.19 |
| | ±SD | 0.66 | 0.01 | 0.31 | 2.59 | 0.01 | 2.95 | 0.64 | 0.05 | 0.11 | 0.01 | 0.02 | 0.14 | 0.00 | 5.22 |
| Aitoloakarnania | Fir | 29.10 | 0.66 | 3.34 | 24.73 | 0.62 | 3.50 | 110.30 | 5.20 | 0.77 | 0.14 | 0.65 | 3.20 | 0.99 | 183.20 |
| Aitoloakarnania | Fir | 18.90 | 0.56 | 5.91 | 24.80 | 0.57 | 7.24 | 46.63 | 2.54 | 0.44 | 0.21 | 0.58 | 1.15 | 0.72 | 110.25 |
| Aitoloakarnania | Fir | 54.56 | 0.08 | 3.43 | 20.59 | 0.95 | 4.84 | 135.36 | 7.26 | 0.78 | 0.08 | 0.14 | 23.76 | 1.18 | 252.93 |
| | Average | 34.19 | 0.43 | 4.23 | 23.37 | 0.72 | 5.19 | 97.43 | 5.00 | 0.66 | 0.18 | 0.46 | 9.37 | 0.96 | 182.13 |
| | ±SD | 18.37 | 0.31 | 1.46 | 2.41 | 0.21 | 1.90 | 45.74 | 2.37 | 0.20 | 0.05 | 0.28 | 12.50 | 0.23 | 71.34 |
| Zagorochoria | Forest flowers | 9.15 | 0.06 | 4.72 | 59.88 | 0.65 | 2.81 | 82.72 | 5.54 | 0.35 | 0.07 | 0.28 | 9.58 | 3.52 | 179.34 |
| Zagorochoria | Forest flowers | 8.30 | 0.05 | 4.47 | 56.71 | 0.64 | 2.63 | 80.70 | 5.32 | 0.57 | 0.05 | 0.19 | 8.84 | 3.45 | 171.90 |
| | Average | 8.73 | 0.06 | 4.59 | 58.30 | 0.64 | 2.72 | 81.71 | 5.43 | 0.46 | 0.06 | 0.24 | 9.21 | 3.49 | 175.62 |
| | ±SD | 0.60 | 0.01 | 0.18 | 2.24 | 0.01 | 0.13 | 1.43 | 0.16 | 0.15 | 0.01 | 0.06 | 0.52 | 0.05 | 5.26 |
| | LOD (mg kg$^{-1}$) | 0.44 | 0.08 | 0.003 | 0.03 | 0.11 | 0.01 | 0.0014 | 0.08 | 0.07 | 0.08 | 0.04 | 0.0011 | 0.0032 | |
| | LOQ (mg kg$^{-1}$) | 1.34 | 0.26 | 0.01 | 0.10 | 0.35 | 003 | 0.0046 | 0.24 | 0.21 | 0.26 | 0.14 | 0.0036 | 0.0106 | |

Each sample was analyzed in triplicate ($n = 3$). LOD: limit of detection. LOQ: limit of quantification.

Furthermore, Ca and Mg content values (mg kg$^{-1}$) in asfaka honeys was significantly lower than that reported for Spanish rosemary, heather, orange blossom, and eucalyptus honeys [13]; Hellenic blossom honeys [15]; Moroccan thyme honey [16]; Egyptian clover honey [5]; or honeydew honeys from Poland [14]. On the contrary, a higher content of Ca was reported for Uruguayan and Brazilian honeys (range of 64.45 ± 25.50–77.60 ± 28.81 and 338.7 ± 14.61 mg kg$^{-1}$, respectively) [2,3].

Furthermore, the content of Al, Mg, and Mn in the 3rd flower honey sample from Aitoloakarnania is significantly higher compared to the other flower or blossom honey samples, but within the range for fir honey samples. This is probably owed to the contribution of honeydew elements during the harvesting of flower honey in the greater area of Aitoloakarnania.

The content (mg kg$^{-1}$) of Fe and Zn in flower honeys from Aitoloakarnania and Samos Island are in agreement with the results reported by Berriel et al. [2] involving prairie, eucalyptus and native woody vegetation honeys from different geographical zones in Uruguay. The higher Zn content was recorded for forest flowers honeys from Zagorochoria followed by orange blossom honeys from Lakonia (Table 1). The content of Zn in Swiss rape honeys (ca. 0.69 ± 0.09) is in agreement with present results involving asfaka honeys from Aitoloakarnania and flower honeys from Samos Island (0.68 ± 0.04 and 0.68 ± 0.21 mg kg$^{-1}$, respectively).

However, the content of Ni in the present study was significantly higher in fir honeys compared to the other honey types analyzed (Table 1). Other researchers have reported lower Ni content values (range of ca. 0.03–0.06 mg kg$^{-1}$) for Swiss acacia, chestnut, dandelion, lime, and rape honeys [4]. What is remarkable is that the Ni content values for Swiss rhododendron and mixed blossom honeys (0.15 and 0.10 mg kg$^{-1}$, respectively) are in agreement with present results involving orange blossom honeys (Table 1). Swiss fir honeys showed a higher Ni content (ca. 1.57 mg kg$^{-1}$) compared to present results [4].

3.1.2. Minor Minerals

Some typical information about the minor mineral content data (mg kg$^{-1}$) that are not listed in the Table 1 follow the text sequence: Barium was identified in one flower honey sample (No. 4) from Samos Island at 0.33 mg kg$^{-1}$. In both forest flowers honey samples from Zagorochoria, Ba was recorded at 0.21 and 0.18 mg kg$^{-1}$, respectively. Fir honey sample (No. 1) contained Ba at 0.12 mg kg$^{-1}$. In asfaka and orange blossom honeys Ba was found at <0.06 mg kg$^{-1}$. Finally, a cotton honey sample from Larissa contained Ba in amounts of <0.17 mg kg$^{-1}$. These content values of Ba are in agreement with a previous work dealing with Moroccan, Spanish, Egyptian and Hellenic thyme honeys [16].

Similar content values were obtained for Cd, in which its content was <0.05 mg kg$^{-1}$ in asfaka, cotton, forest flowers and orange blossom honeys. Cadmium was identified in an amount of <0.05 mg kg$^{-1}$ only in 1 fir honey sample from Aitoloakarnania, whereas it was not detected in the other honey types investigated. The maximum level of 0.1 mg kg$^{-1}$ for cadmium in honey has been suggested by the EU [18]. All honey samples analyzed conform to this upper limit and are slightly higher compared to the results reported for Swiss blossom and honeydew honeys [4]. The low Cd content values have been also confirmed in a previous work dealing with Hellenic pine, thyme, wild thyme, citrus, multifloral, and mixed citrus with erica honeys [5,15]. Chromium was identified only in 3 flower honey samples from Samos Island (Nos. 1, 5 and 7) at 0.18, 0.39 and 0.18 mg kg$^{-1}$, respectively. In the other honey types, Cr was either not detectable (all flower honeys from Aitoloakarnania, 2 fir honeys from Aitoloakarnania (samples 1 & 2), and flower honey samples Nos. 2, 3, 4, 6 from Samos Island), or <0.12 mg kg$^{-1}$ (asfaka, fir, cotton, forest flowers, and orange blossom honeys). These values agree with those reported for European and Mediterranean honeys [4,16].

Cobaltum content was <0.03 mg kg$^{-1}$ in asfaka, cotton, forest flowers and orange blossom honeys, while a value of <0.11 mg kg$^{-1}$ was recorded in a 1 fir honey sample from Aitoloakarnania. In flower honeys from Aitoloakarnania and Samos Island, cobaltum was absent. Similarly, mercury content was <0.03 mg kg$^{-1}$ in asfaka, cotton, forest flowers and orange blossom honeys, while a value of

1.25 mg kg$^{-1}$ was recorded in 1 fir honey sample (No. 2) from Aitoloakarnania. In flower honeys from Aitoloakarnania and Samos Island, mercury was also absent.

Molybdenum was identified in 4 flower honey samples from Samos Island (Nos. 1, 2, 3, and 6) at 0.27, 0.16, 0.21 and 0.22 mg kg$^{-1}$, respectively, whereas in fir honey samples from Aitoloakarnania only in sample No. 2 at 0.3 mg kg$^{-1}$. Among the rest, honey types/samples were either not detectable (all flower honeys from Aitoloakarnania, 3 flower honeys (Nos. 4, 5 and 7) from Samos Island) and 1 fir honey sample (No. 1) from Aitoloakarnania, or <0.08 mg kg$^{-1}$.

The same holds true for Se, which was identified in higher amounts in the 5 flower honey samples from Samos Island (Nos. 1, 2, 3, 5 and 7) at 0.41, 0.42, 0.41, 0.39 and 0.64 mg kg$^{-1}$, respectively, whereas in the rest, honey types/samples were either not detectable (all flower honeys from Aitoloakarnania, 2 flower honeys (Nos. 4, 6 from Samos Island) and 2 fir honey samples (Nos. 1 and 2 from Aitoloakarnania) or was <0.12 mg kg$^{-1}$. In previous studies dealing with a similar topic, Se content recorded, in agreement with present results, to be higher in content in Hellenic pine and thyme honeys (0.42 ± 0.26 and 0.380.42 mg kg$^{-1}$, respectively) [15] compared to Moroccan (0.04 ± 0.06 mg kg$^{-1}$), Spanish (0.16 ± 0.12 mg kg$^{-1}$) and Egyptian (0.25 ± 0.13 mg kg$^{-1}$) thyme honeys [16].

Silver was not detected in the honey samples of the present work. Beryllium recorded a very low content (<0.06 mg kg$^{-1}$) in asfaka, cotton, forest flowers, orange blossom honeys, whereas this amount was also identified only in 1 fir honey sample (No. 3) from Aitoloakarnania. In flower honeys from Aitoloakarnania and Samos Island it was not detected. These findings are in agreement with the results reported for Hellenic pine and thyme honeys [15].

Titanium was identified in flower honey samples (Nos. 1, 5 and 7) from Samos Island at higher amounts (0.11 mg kg$^{-1}$), whereas in asfaka, cotton, forest flowers, and orange blossom honeys was <0.09 mg kg$^{-1}$. Thallium was identified in low amounts (<0.02 mg kg$^{-1}$) in asfaka, cotton, forest flowers and orange blossom honeys, in agreement with the results reported in previous studies dealing with pine, thyme, multiflower, and orange blossom honeys of the Mediterranean zone [5,16].

Finally, vanadium (V) was not identified in all flower honey samples from Aitoloakarnania and Samos Island along with 2 fir honeys from Aitoloakarnania. Amounts of V < 0.10 mg kg$^{-1}$ were identified in the remaining asfaka, fir, cotton, forest flowers and orange blossom honeys in agreement with previous studies involving Mediterranean pine and thyme honeys [15,16].

### 3.2. Multivariate Statistics

#### 3.2.1. MANOVA

The level of significance and the ability of minerals/total mineral content to be used for the botanical origin differentiation of asfaka, cotton, fir, flower, forest flowers, and orange blossom honeys, was defined by using MANOVA. The independent variables comprised the 13 minerals and total mineral content while botanical origin was taken as the grouping (dependent) variable.

Pillai's Trace = 4.790 (F = 8.759, $p = 0.000 < 0.001$) and Wilks' Lambda = 0.000 (F = 44.369, $p = 0.000 < 0.001$) values showed the existence of a significant multivariable effect of honey minerals and total mineral content on the latter botanical origin. Ten of the 13 minerals and total mineral content were found to be significant ($p < 0.05$) for the botanical origin differentiation of honeys (Table S1).

#### 3.2.2. FA

During the FA, the K1 rule was considered in order to investigate whether the sample size could affect the reliability of analysis. The respective value of Kaiser-Meyer-Olkin (KMO) measure of sampling adequacy was 0.431 < 0.50, indicating that some minor sample issues existed. Based on this observation, the Bartlett's test of sphericity test was also applied. Bartlett's test of sphericity tests the hypothesis that the correlation matrix is an identical matrix, which would indicate that the variables are unrelated and therefore unsuitable for structure detection. Values less than 0.05 of the significance level indicate that a factor analysis may be useful with the set of data collected. Respective values of

Bartlett's test of sphericity were X2 = 299.314, df = 55, $p$ = 0.000, indicating the effectiveness of FA in the group of samples investigated.

The extraction method was PCA. Communalities of the variables used in FA were also considered between the initial (all variables had the value 1.000) and extraction values (Table S2). It should be noted that, if the extraction values were lower than 0.3 then problems in FA analysis may arise [24]. During the analysis this problem did not arise. FA was carried by application of Varimax with Kaiser Normalization rotation method. It should be defined that the Varimax rotation of the axles is the best rotation method since it defines better the group of variables that are created and guarantees that these variables are not correlated. In addition, the possibility of co-linearity between the variables is eliminated. The structure matrix consisted of 4 principal components (eigenvalue greater than 1).

More specifically, PCA 1 (eigenvalue of 4.476) explained 40.687% of total variance and consisted of Mg, Ni, Al, Mn, and TM. PCA 2 (eigenvalue of 2.782) explained 25.291% of total variance and consisted of As, Sb, and Pb. Similarly, PCA3 (eigenvalue of 1.761) explained 16.010% of total variance and consisted of Ca and Zn. Finally, PCA 4 (eigenvalue of 1.331) explained 12.103 % of total variance and consisted of Si (Figure S1). The criterion to categorize the aforementioned variables in the four principal components was the higher absolute value of communalities (Table S3) that built the rotated component matrix. Therefore, the four principal components explained 94.091% of total variance which is a very satisfactory rate for honey botanical origin differentiation.

*3.3. Botanical Origin Differentiation of Hellenic Asfaka, Cotton, Flower, Fir, Forest Flowers and Orange Blossom Honeys Based on Abundant Mineral Content and Linear Discriminant Analysis*

Results showed that 5 discriminant functions (DFs) were formed: (i) DF1:Wilks' Lambda = 0.000, X2 = 157.870, df = 45, $p$ = 0.000, (ii) DF2: Wilks' Lambda = 0.000, X2 = 99.277, df = 32, $p$ = 0.000, (iii) DF3: Wilks' Lambda = 0.004, X2 = 58.925, df = 21, $p$ = 0.000, (iv) DF4: Wilks' Lambda = 0.064, X2 = 28.880, df = 12, $p$ = 0.000, and (v) DF5: Wilks; Lambda = 0.357, X2 = 10.810, df = 5, $p$ = 0.055. It should not be forgotten, that a significant value ($p < 0.05$) of Wilks' Lambda shows that the discriminant function is basic for the differentiation of the investigated groups. It is clear, then, that DF5 could not contribute to the botanical origin differentiation of asfaka, cotton, flower, forest flowers and orange blossom honeys.

The first discriminant function showed the highest eigenvalue (264.153) and canonical correlation of 0.998, whereas accounted for 79.4% of total variance; the second discriminant function showed an eigenvalue of 45.665, canonical correlation of 0.989 and accounted for 13.7% of total variance; the third discriminant function showed an eigenvalue of 16.486, canonical correlation of 0.971 and accounted for 5.0% of total variance; discriminant function 4 showed a lower eigenvalue (4.590), a weaker canonical correlation (0.906) and explained 1.4% of total variance. The non-significant discriminant function 5 showed the lowest eigenvalue (1.800), the weakest canonical correlation (0.802), whereas explained only 0.5% of total variance. All Dfs accounted for 100% of total variance. The unstandardized canonical discriminant functions evaluated at group means, the group centroids, had the following values: (4.289, 5.063), (−2.134, −3.331), (6.982, −1.003), (−8.418, −13.393), (10.336, 4.691), (−35.821, 5.602) for flower honey from Aitoloakarnania, Asfaka honey from Aitoloakarnania, flower honey from Samos Island, orange honey from Lakonia, fir honey from Aitoloakarnania, and forest flowers honey from Zagorochoria, respectively.

In Figure S2 it is shown that honey types are satisfactorily differentiated. The first discriminant function clearly differentiates flower honeys from Aitoloakarnania, flower honeys from Samos Island and fir honeys from Aitoloakarnania, while the second discriminant function differentiates orange blossom honeys from Lakonia and cotton honeys from Larissa. The overall correct classification rate was 100% using the original and 78.9% the cross validation method. The botanical classification rate for honey samples based on the cross validation method followed the sequence: Flower honeys from Aitoloakarnania (33.3%), asfaka honeys from Aitoloakarnania (100%), flower honeys from Samos Island (100%), orange blossom honeys from Lakonia (100%), fir honeys from Aitoloakarnania (33.3%), and cotton honeys from Larissa (100%). Regarding the low prediction rates of flower and for honeys from Aitoloakarnania, it should be stressed that from the 3 samples studied, in both cases, only 1 sample was correctly classified. In particular, of the flower honey samples from Aitoloakarnania 1 sample was correctly allocated in flower honey group, while the other two were allocated in fir honey group. Regarding fir honey from Aitoloakarnania, of the 3 samples studied, 1 sample was correctly allocated in the fir honey group, while the other two were allocated in the flower honey group from Aitoloakarnania.

### 3.4. Botanical Origin Differentiation of Hellenic Asfaka, Cotton, Flower, Fir, Forest Flowers and Orange Blossom Honeys Based on Abundant Mineral Content and Stepwise Discriminant Analysis

To investigate further, whether the correct prediction rates of honeys according to botanical origin could be improved by the selection of the most potential minerals, stepwise discriminant analysis was applied. Results showed that 5 discriminant functions (DFs) were formed: (i) DF1:Wilks' Lambda = 0.000, X2 = 178.081, df = 35, $p$ = 0.000, (ii) DF2: Wilks' Lambda = 0.000, X2 = 92.667, df = 24, $p$ = 0.000, (iii) DF3: Wilks' Lambda = 0.012, X2 = 51.122, df = 15, $p$ =0.000, (iv) DF4: Wilks' Lambda = 0.086, X2 = 28.270, df = 8, $p$ = 0.000, and (v) DF5: Wilks' Lambda = 0.397, X2 = 10.637, df = 3, $p$ = 0.014.

The first discriminant function showed the higher eigenvalue (1680,290) and canonical correlation of 1.000, whereas it accounted for 97.3% of total variance; the second discriminant function showed an eigenvalue of 36.061, canonical correlation of 0.986 and accounted for 2.1% of total variance; the third discriminant function showed an eigenvalue of 6.295, canonical correlation of 0.929 and accounted for 0.4% of total variance; discriminant function 4 showed a lower eigenvalue (3.633), a weaker canonical correlation (0.886) and explained 0.2% of total variance; discriminant function 5 showed the lowest eigenvalue (1.522), the weakest canonical correlation (0.777, whereas explained only 0.1% of total variance. All Dfs accounted for 100% of total variance. The unstandardized canonical discriminant functions evaluated at group means, the group centroids, had the following values: (15.597, 4.858), (−1.623, −0.323), (8.421, −5.520), (−20,929, −2.154), (38.207, 6.755), (−87.627, 4.378) for flower honey from Aitoloakarnania, Asfaka honey from Aitoloakarnania, flower honey from Samos Island, orange honey from Lakonia, fir honey from Aitoloakarnania, and forest flowers honey from Zagorochoria, respectively.

Stepwise discriminant analysis in Figure 1 shows that all honey types are perfectly differentiated. The first discriminant function clearly differentiates flower honeys from Aitoloakarnania, flower honeys from Samos Island and fir honeys from Aitoloakarnania, while the second discriminant function differentiates asfaka honeys from Aitoloakarnania, orange blossom honeys from Lakonia and cotton honeys from Larissa. The overall correct classification rate was 100% using the original and 100% the cross validation method. In all cases, the botanical classification rate of the honey types studied was 100%.

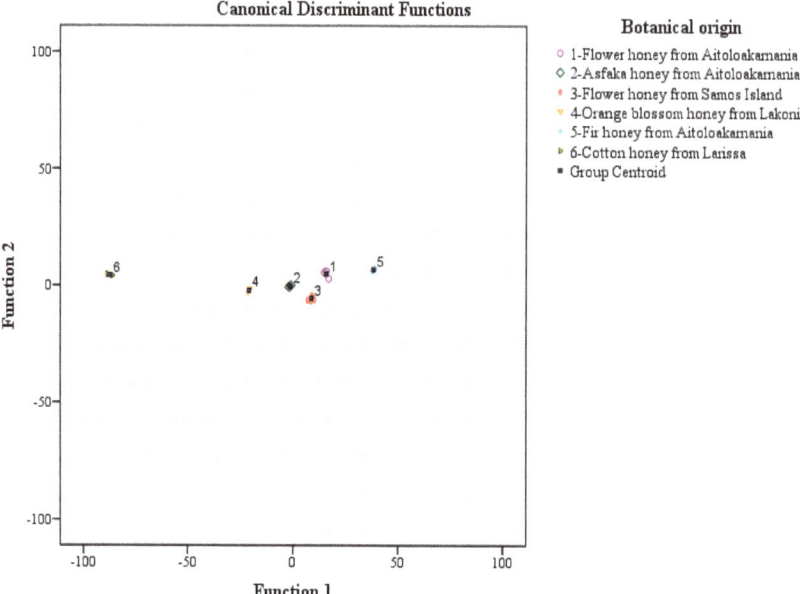

**Figure 1.** Botanical origin differentiation of Hellenic asfaka, cotton, flower, fir, forest flowers and orange blossom honeys based on abundant mineral and total mineral contents and stepwise discriminant analysis.

## 4. Overview of the Chemometrics Affinity to the Botanical Origin Differentiation of Asfaka, Cotton, Fir, Flower, Forest Flowers and Orange Blossom Honeys

The different chemometric techniques that were applied to the mineral content analysis data provided exhaustive information about their contribution to the botanical origin differentiation of asfaka, cotton, fir, flower, forest flowers and orange blossom honeys. MANOVA highlighted significant minerals ($p < 0.05$) that could contribute to the botanical origin differentiation of the aforementioned honeys. FA revealed that mineral content data could explain by a high degree (ca. 94.10%) the total variance of variables in the multi-dimensional space. Despite the minor sample size issues that existed during FA analysis by application of the KMO rule, the significant values of Bartlett's test of sphericity showed the effectiveness of FA in the explanation of the total variance in the group of samples investigated, based on mineral content analysis data.

However, at this point some useful remarks should be recorded: According to the statistical theory, if the sample size is increased, it is reasonable to have different significant variables because the size of samples affects the confidence level [25]. In addition, it is not always a surety that by increasing the sample size, the correct classification rates of the investigated group of objects, will increase. Finally, if there were any computational problems owed to the sample size, the SPSS program would not run any analysis. Therefore, quality analysis criteria such as that of the preset study should be taken into account in cases where sample collection/sample size is, somehow, limited.

Regarding the application of LDA and SDA for the complete classification of honeys according to botanical origin based on mineral content analysis data (correct classification rates of 75% and 100%, respectively), the variables contributing the most to the overall discrimination rate were those with the highest absolute pooled-within correlation values between discriminating variables and standardized canonical discriminant functions (Table 2). These minerals may be, thus, considered as potential factors of the developed discrimination model. In that sense, for the LDA analysis these were: As (DF4), Ca (DF4), Pb (DF4), Sb (DF4), Ni (DF5), Mg (DF5), Al (DF5), Zn (DF5), and Mn (DF5). Similarly, for the

SDA analysis the potential minerals were: As (DF3), Mg (DF5), Al (DF5), Si (DF5), Mn (DF5), Zn (DF5), and Ca (DF5).

Table 2. Contribution of minerals to the discriminant function matrix during LDA and SDA.

| | Structure Matrix–LDA | | | | |
|---|---|---|---|---|---|
| Minerals | Function | | | | |
| | 1 | 2 | 3 | 4 | 5 |
| As | 0.095 | 0.108 | −0.186 | 0.476 * | −0.087 |
| Ca | −0.083 | 0.053 | −0.128 | 0.421 * | 0.403 |
| Pb | 0.063 | 0.037 | −0.243 | 0.397 * | −0.042 |
| Sb | 0.047 | 0.099 | −0.050 | 0.275 * | −0.027 |
| TM [a] | 0.045 | 0.145 | 0.056 | 0.045 | 0.806 * |
| Ni | −0.003 | 0.111 | 0.072 | −0.231 | 0.729 * |
| Mg | −0.001 | 0.122 | −0.004 | −0.005 | 0.697 * |
| Al | 0.020 | 0.122 | 0.201 | −0.181 | 0.608 * |
| Zn | −0.215 | −0.004 | 0.194 | 0.301 | 0.607 * |
| Mn | −0.037 | 0.164 | 0.156 | −0.087 | 0.582 * |
| Si [a] | 0.336 | 0.081 | 0.047 | −0.147 | 0.559 * |
| | Structure Matrix–SDA | | | | |
| Minerals | Function | | | | |
| | 1 | 2 | 3 | 4 | 5 |
| Sb [b] | 0.165 | 0.104 | 0.650 * | −0.081 | −0.151 |
| As | 0.033 | −0.106 | 0.591 * | 0.163 | −0.203 |
| Pb [b] | 0.294 | −0.171 | −0.350 | 0.620 * | −0.127 |
| TM [b] | 0.020 | 0.105 | 0.307 | 0.016 | 0.816 * |
| Mg | 0.002 | 0.087 | 0.264 | −0.074 | 0.734 * |
| Al | 0.015 | 0.188 | 0.043 | −0.060 | 0.632 * |
| Si | 0.002 | −0.145 | 0.226 | −0.139 | 0.600 * |
| Mn | −0.008 | 0.222 | 0.172 | −0.067 | 0.598 * |
| Ni [b] | 0.219 | 0.176 | 0.172 | 0.028 | 0.588 * |
| Zn | −0.080 | 0.209 | −0.103 | 0.501 | 0.582 * |
| Ca | −0.035 | −0.002 | 0.352 | 0.253 | 0.368 * |

Pooled within-groups correlations between discriminating variables and standardized canonical discriminant functions. Variables ordered by absolute size of correlation within function. * Largest absolute correlation between each variable and any discriminant function. [a,b] This variable was not used in the analysis.

Fernández-Torres et al. [13] classified eucalyptus, heather, orange blossom, and rosemary honeys from different regions in Spain by applying LDA analysis to Zn, P, B, Mn, Mg, Cu, Ca, Sr, Ba, Na, and K contents (mg kg$^{-1}$). The overall classification rate was 97% based on the cross validation method.

Bogdanov et al. [4] classified acacia, chestnut, dandelion, lime, rape, rhododendron, fir, mountain blossom and mixed blossom honeys according to botanical origin based on Cd, Pb, Cr, Mn, Fe, Ni, Cu, and Zn content (mg kg$^{-1}$) and LDA analysis. The overall correct classification rate was 76%. However, the individual classification rates were 100%, 100%, 100, 88%, and 83% for acacia chestnut, fir, dandelion, and rape honeys, respectively.

Chudzinska and Baralkiewicz [14], classified Polish rape, buckwheat and honeydew honeys based on K and Mn contents (mg kg$^{-1}$) in combination with LDA analysis. These authors reported classification rates of 100%, 83% and 70% for rape, buckwheat and honeydew honeys, respectively.

Berriel et al. [2] classified prairie, eucalyptus and native woody vegetation honeys from different geographical zones in Uruguay using K, Ca, Na, Mf, Fe, Mn, Zn and Cu and discriminant analysis. The overall correct classification rates based on the original and cross validation methods were 91% and 65%, respectively.

## 5. Conclusions

Mineral content analysis implemented with different chemometric tools may accurately assist in the botanical origin differentiation of different varieties of beekeepers' honey. The present work, apart from the mineral content characterization of asfaka, cotton, fir, flower, forest flowers and orange blossom honeys, assists in the quality control analysis of honey obtained directly from beekeepers by providing information on specific minerals, total mineral content, or the level of toxic metals such as Pb, Cd, and Cr. It is mandatory for honey consumers to know the exact quality of honey they buy, as their rights, health, and diet casualties, are well protected. Based on this ethical good, and given the fact that Pb content in most of the samples was higher than the regulated upper level, there is a great tendency for further improvement of the beekeepers' practices for honey production.

**Supplementary Materials:** The following are available online at http://www.mdpi.com/2304-8158/8/6/210/s1, Table S1: Level of significance of individual minerals and total mineral content on honey botanical origin as defined by MANOVA, Table S2: Minerals and total mineral content used to build the rotated component matrix[a] along with communalities during extraction (Extraction method: Principal component analysis. Rotation method: Varimax with Kaiser Normalization. Rotation converged in 5 iterations. TM: total minerals. Initial communalities: all variables had the value of 1.000), Table S3: Classification ability of the SDA analysis model used for the botanical origin differentiation of asfaka, cotton, fir, flower, forest flowers, and orange blossom honeys, Figure S1: Factor analysis highlighting the principal components (structure matrix) used for the botanical origin differentiation of Hellenic asfaka, cotton, fir, flower, forest flowers and orange blossom honeys based on abundant mineral and total mineral contents. Extraction method: Principal component analysis. Rotation method: Varimax with Kaiser Normalization. Figure S1 shows only the first 3 components in dimensional space X, Y, Z. TM: total mineral content, Figure S2: Botanical origin differentiation of Hellenic asfaka, cotton, flower, fir, forest flowers and orange blossom honeys based on abundant mineral and total mineral contents and linear discriminant analysis.

**Author Contributions:** Conceptualization, I.K.K; methodology, A.P.L., C.P., I.K.K.; software, C.P.; validation, A.P.L., C.P. and I.K.K.; formal analysis, A.P.L., I.K.K.; investigation, I.K.K., A.B.; resources, C.P.; data curation, A.P.L., I.K.K.; writing—original draft preparation, I.K.K.; writing—review and editing, I.K.K.; visualization, I.K.K.; supervision, I.K.K.; project administration, I.K.K.

**Funding:** This research received no external funding.

**Acknowledgments:** The authors are grateful to local beekeepers from Aitolokarnania, Lakonia, Larissa, Samos Island, Zagorochoria, for the donation of honey samples. M.G. Kontominas is acknowledged for the collection of asfaka, fir and flower honey samples from Aitoloakarnania.

**Conflicts of Interest:** The authors declare no conflict of interest.

## References

1. White, J.W. Physical characteristics of honey. In *Honey, A Comprehensive Survey*; Crane, E., Ed.; Heinemann: London, UK, 1975; pp. 207–239.
2. Berriel, V.; Barreto, P.; Perdomo, C. Characterisation of Uruguayan honeys by multi-elemental analyses as a basis to assess their geographical origin. *Foods* **2019**, *8*, 24. [CrossRef] [PubMed]
3. Kadri, S.M.; Zaluski, R.; Pereira Lima, G.P.; Mazzafera, P.; De Oliveira Orsi, R. Characterization of Coffea arabica monofloral honey from Espírito Santo, Brazil. *Food Chem.* **2016**, *203*, 252–257. [CrossRef] [PubMed]
4. Bogdanov, S.; Haldimann, M.; Luginbühl, W.; Gllmann, P. Minerals in honey: Environmental, geographical and botanical aspects. *J. Apic. Res.* **2007**, *46*, 269–275. [CrossRef]
5. Karabagias, I.K.; Louppis, A.; Kontakos, S.; Drouza, C.; Papastephanou, C. Characterization and botanical differentiation of monofloral and multifloral honeys produced in Cyprus, Greece and Egypt using physicochemical parameter analysis and mineral content, in conjunction with supervised statistical techniques. *J. Anal. Meth. Chem.* **2018**, *2018*, 10. [CrossRef] [PubMed]
6. Von der Ohe, W.; Persano Oddo, L.; Piana, M.L.; Morlot, M.; Martin, P. Harmonized methods of melissopalynology. *Apidologie* **2004**, *35*, S18–S25. [CrossRef]
7. Council Directive 2001/110/EC Relating to Honey. Available online: https://www.ecolex.org/details/legislation/council-directive-2001110ec-relating-to-honey-lex-faoc037441/ (accessed on 26 May 2019).
8. Corbella, E.; Cozzolino, D. Classification of the floral origin of Uruguayan honeys by chemical and physical characteristics combined with chemometrics. *LWT-Food Sci. Technol.* **2006**, *39*, 534–539. [CrossRef]

9. Karabagias, I.K.; Halatsi, E.Z.; Karabournioti, S.; Kontakos, S.; Kontominas, M.G. Impact of physicochemical parameters, pollen grains and phenolic compounds for the correct geographical differentiation of fir honeys produced in Greece as assessed by multivariate analyses. *Int. J. Food Prop.* **2017**, *20*, S520–S533. [CrossRef]
10. Stanimirova, I.; Üstun, B.; Cajka, T.; Riddelova, K.; Hajslova, J.; Buydens, L.M.C.; Walczak, B. Tracing the geographical origin of honeys based on volatile compounds profiles assessment using pattern recognition techniques. *Food Chem.* **2007**, *118*, 171–176. [CrossRef]
11. Sergiel, I.; Pohl, P.; Biesaga, M. Characterisation of honeys according to their content of phenolic compounds using high performance liquid chromatography/tandem mass spectrometry. *Food Chem.* **2014**, *145*, 404–408. [CrossRef] [PubMed]
12. Rückriemen, J.; Henle, T. Pilot study on the discrimination of commercial Leptospermum honeys from New Zealand and Australia by HPLC–MS/MS analysis. *Eur. Food Res. Technol.* **2018**, *244*, 1203–1209. [CrossRef]
13. Fernández-Torres, R.; Pérez-Bernal, J.L.; Bello-López, M.A.; Callejón-Mochón, M.; Jiménez-Sánchez, J.C.; Guiraúm-Pérez, A. Mineral content and botanical origin of Spanish honeys. *Talanta* **2005**, *65*, 686–691. [CrossRef]
14. Chudzinska, M.; Baralkiewicz, D. Application of ICP-MS method of determination of 15 elements in honey with chemometric approach for the verification of their authenticity. *Food Chem. Toxicol.* **2011**, *49*, 2741–2749. [CrossRef] [PubMed]
15. Karabagias, I.K.; Louppis, A.P.; Kontakos, S.; Papastephanou, C.; Kontominas, M.G. Characterization and geographical discrimination of Greek pine and thyme honeys based on their mineral content, using chemometrics. *Eur. Food Res. Technol.* **2017**, *243*, 101–113. [CrossRef]
16. Karabagias, I.K.; Louppis, A.P.; Karabournioti, S.; Kontakos, S.; Papastephanou, C.; Kontominas, M.G. Characterization and classification of commercial thyme honeys produced in specific Mediterranean countries according to geographical origin, using physicochemical parameter values and mineral content in combination with chemometrics. *Eur. Food Res. Technol.* **2017**, *243*, 889–900. [CrossRef]
17. Karabagias, I.K.; Vlasiou, M.; Kontakos, S.; Drouza, C.; Kontominas, M.G.; Keramidas, A.D. Geographical discrimination of pine and fir honeys using multivariate analyses of major and minor honey components identified by $^1$H NMR and HPLC along with physicochemical data. *Eur. Food Res. Technol.* **2018**, *244*, 1249–1259. [CrossRef]
18. European Commission. Commission Regulation (EU) No. 2015/1005 of 25 June 2015 amending Regulation (EC) No 1881/2006 as regards maximum levels of lead in certain foodstuffs. *Off. J. Eur. Union* **2015**, *161*, 9–13.
19. Beretta, G.; Caneva, E.; Regazzoni, L.; Bakhtyari, N.G.; Facino, R.M. A solid-phase extraction procedure coupled to $^1$H NMR, with chemometric analysis, to seek reliable markers of the botanical origin of honey. *Anal. Chim. Acta* **2008**, *620*, 176–182. [CrossRef] [PubMed]
20. Warne, R.T. A primer on multivariate analysis of variance (MANOVA) for behavioral scientists. *Pract. Assess. Res. Eval.* **2014**, *19*, 1–10.
21. Huberty, C.J.; Olejnik, S. *Applied MANOVA and Discriminant Analysis*, 2nd ed.; Wiley: Hoboken NJ, USA, 2006; pp. 300–301.
22. Jolliffe, I.T. *Principal Component Analysis, Series: Springer Series in Statistics*, 2nd ed.; Springer: New York, NY, USA, 2002; p. 487.
23. IBM Knowledge Center. Available online: https://www.ibm.com/support/knowledgecenter/en/SSLVMB_23.0.0/spss/tutorials/discrim_telco_stepwise.html (accessed on 30 April 2019).
24. Karabagias, I.K.K. Seeking of reliable markers related to Greek nectar honey geographical and botanical origin identification based on sugar profile by HPLC-RI and electro-chemical parameters using multivariate statistics. *Eur. Food Res. Technol.* **2019**, *245*, 805–816. [CrossRef]
25. Field, A. *Discovering Statistics Using SPSS*, 3rd ed.; Sage Publications Ltd.: London, UK, 2009; p. 384.

© 2019 by the authors. Licensee MDPI, Basel, Switzerland. This article is an open access article distributed under the terms and conditions of the Creative Commons Attribution (CC BY) license (http://creativecommons.org/licenses/by/4.0/).

MDPI  
St. Alban-Anlage 66  
4052 Basel  
Switzerland  
Tel. +41 61 683 77 34  
Fax +41 61 302 89 18  
www.mdpi.com

*Foods* Editorial Office  
E-mail: foods@mdpi.com  
www.mdpi.com/journal/foods

www.ingramcontent.com/pod-product-compliance
Lightning Source LLC
LaVergne TN
LVHW070606100526
838202LV00012B/575